Basic Carrier Telephony

REVISED THIRD EDITION

David Talley
Telecommunications Consultant

A DIVISION OF HAYDEN PUBLISHING COMPANY
HASBROUCK HEIGHTS, NEW JERSEY / BERKELEY, CALIFORNIA

Library of Congress Cataloging in Publication Data

Talley, David.
Basic carrier telephony.

Includes index.
1. Telephone, Carrier current. I. Title.
TK6421.T3 1977 621.387'8 77-2828
ISBN 0-8104-5848-9

Printed in the United States of America

PRINTING

85 YEAR

Preface

This third revised edition of *Basic Carrier Telephony* includes simplified and understandable presentations of telephone carrier or multiplexing techniques, as well as various technical developments since the original publication of this book. A new section describing broadband carrier systems and their applications has been included. Various additions and changes in the text and drawings have been made to indicate state-of-the-art developments, especially in PCM carrier systems. This latest revised text should be particularly helpful to those responsible for the management, engineering, operation, and maintenance of telecommunication services.

The multiplexing of telephone circuits so that one pair of wires, for example, can carry several conversations was, until recent years, largely confined to the intertoll telephone network of cables and microwave radio facilities. As a result, relatively few engineers, technicians, and maintenance personnel outside of the long lines network understood or were concerned with the intricacies of carrier equipment and its operations. This concept was changed by the many advances in the communications art that resulted in the rapid expansion of the multiplex technique to the short-haul toll and local telephone network, satellite and space communications, and various commercial and military communication applications.

Concise and descriptive narration, with carefully related illustrations, are utilized in this book. Such treatment and the relative absence of mathematics afford the reader—whether he or she be an engineer, technician or student—a better understanding of the principles and applications of carrier telephony and its place in the overall communications picture.

DAVID TALLEY

Contents

Growth of Telephone Service

Telephone service is something that most of us take for granted. We are so accustomed to dialing the number of a friend in the next block, nearby town, or a distant city that we do not realize the vast amount of instruments, central-office switching equipment, overhead wires, cables, and radio circuits that make possible our conversations.

There were over 360 million telephones in the world in 1975 compared to 183 million in 1965, an increase of 97%. The following table is a comparison of the growth of telephone service in various countries during the past decade.

*World's 10 Leading Telephone Countries**

Country	1975	1965	Increase
United States	143,972,000	88,793,000	62%
Japan	41,904,960	12,250,841	242%
United Kingdom	20,342,457	9,960,000	104%
West Germany	18,767,033	8,168,188	130%
U.S.S.R.	15,782,000	7,100,000	122%
Italy	13,695,006	5,528,751	148%
Canada	12,454,331	7,019,374	77%
France	12,405,000	5,703,878	117%
Spain	7,042,968	2,526,843	179%
Sweden	5,178,082	2,386,925	117%

* "The World's Telephones," 1975 AT&T Co.

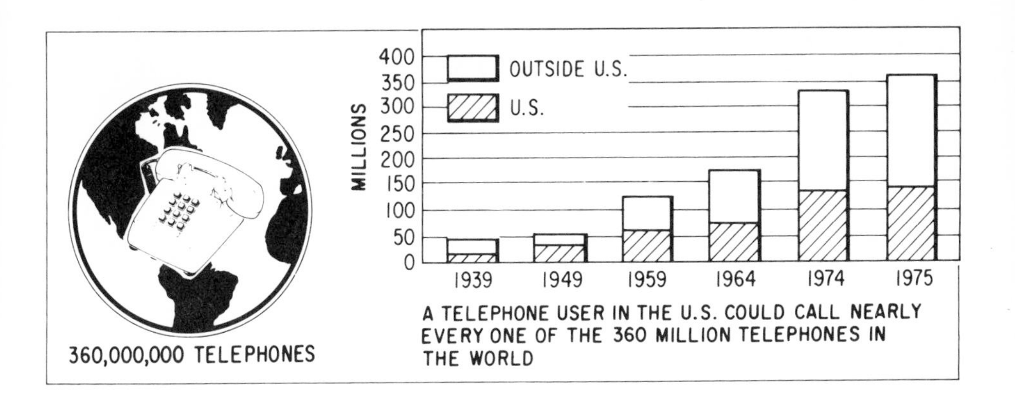

The increase in the population of the United States, the rapid expansion of suburban and rural areas, especially around cities, and the continued economic growth of the country since World War II have contributed to the current high demands for telephone service. Not only were more telephones needed to serve the larger population, but there has been an even greater demand for improved long distance and toll services.

Telephone Statistics

Close to one-half of the world's telephones are in the United States at the present time. The increasing demand for more telephones and greater long distance connections is of particular significance to our study of the *basics of carrier telephony.* Some pertinent data which appeared in recent annual reports of the American Telephone and Telegraph Company (AT&T) are listed herein. This organization through its 23 principal telephone operating companies comprising the Bell System operates about 80% of the telephones in the United States. Over 13,500,000 additional telephones are served by the General Telephone System, the second largest telephone organization, and the remainder are controlled by over 1,100 small Independent Telephone companies scattered all over the country. All of these telephone companies connect with the long distance and toll circuits of AT&T so that any telephone in the country can be connected to any other one whether served by the Bell System, the General Telephone System, or any of the other Telephone Systems and the many Independent Telephone companies.

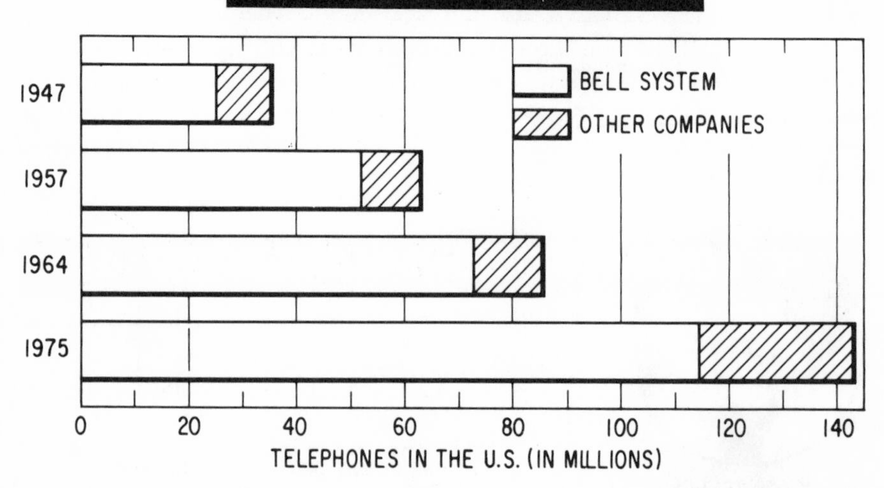

Each telephone company serves a specific locality or territory, usually under a franchise from the state or municipal government. The rates and operations of these telephone companies are under the administrative control of the respective state's regulatory commissions. The Federal Communications Commission (FCC) has jurisdiction over the interstate operation and rate structures of telephone companies.

Plant Investments and Revenues

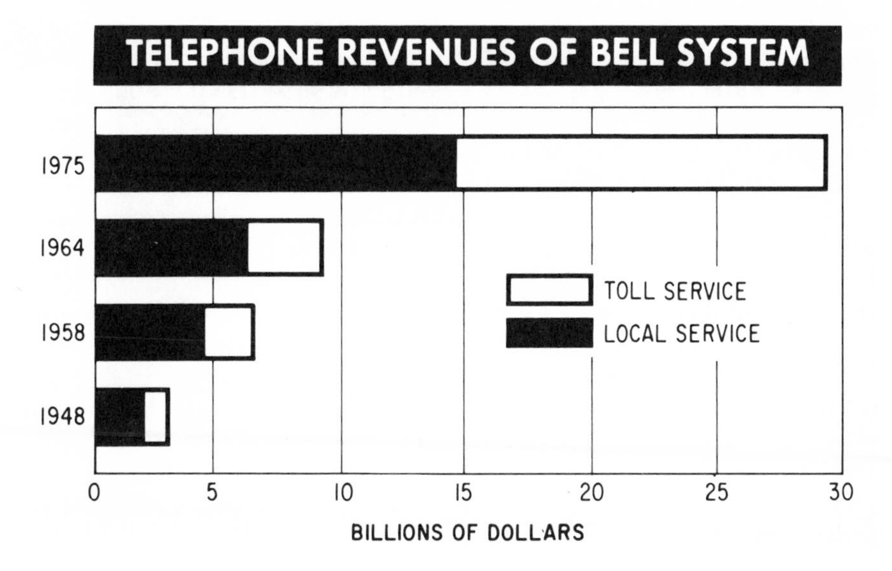

To satisfactorily serve the telephone requirements of the American people requires a great amount of plant and equipment. These facilities cost a great deal of money. The estimates of plant investments and telephone revenues of the Bell System are of interest in this connection. Note that the money received from toll and long distance calls is about 49% of the indicated telephone revenues. The economic importance of adequate and satisfactory long distance communications to interconnect the cities, towns, and other communities of the United States is self-evident.

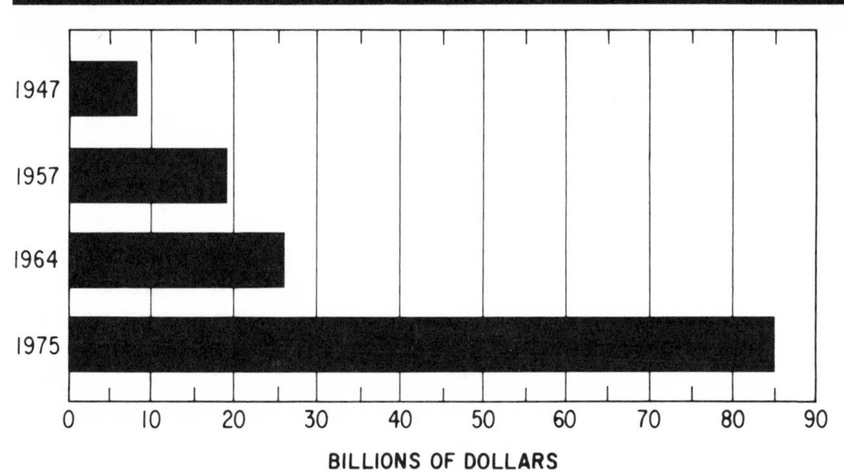

Telephone Plant and Central Offices

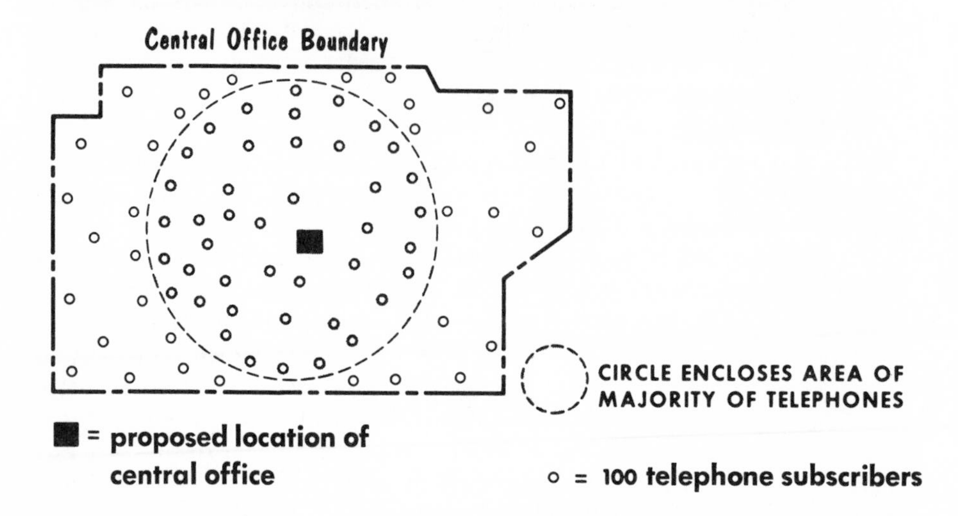

In cities or other large urban and suburban areas, the local service telephone system can be considered to be divided mainly into the subscribers' or customer plant and the trunk plant. The interface between them is the central office or telephone exchange with its switching, transmission, and control equipment.

To serve the nearly 114 million telephones of the Bell System in 1975 required almost 16,500 central offices. Another 13,000 or so exchanges serve the more than 28 million telephones of Independent companies. Their trunks interconnect with the Bell System so that calls can be completed anywhere in the nation. Over 99.9% of all central offices are now automatic or dial-operated. In large cities, several dial central offices may be located in one central building because of the great number of telephones served.

Central offices connect to the telephones in the homes and business places of subscribers by overhead distribution and underground multiconductor (many wires) cables. A pair of wires, commonly called the *tip* and *ring* conductors, is required to connect individual telephone stations to the associated equipment in the telephone exchange. It is customary to establish the central office in the contemplated *wire center* of the telephone subscribers; that is, at the center of a circle whose circumference encloses the area in which most of the telephones will be located. This permits the most economical arrangement of the overhead distribution wires and cables as well as the underground cables which comprise the *Exchange or Outside Plant.*

Interoffice Trunks and Toll Network

Local or end central offices are connected together by *trunk circuits,* each consisting of a pair of wires. These circuits or interoffice trunks use larger size cable conductors (usually No. 19 gage) than those in subscribers cables. The latter normally have No. 22, No. 24, or No. 26 gage copper wires. Larger size conductors are needed for such trunk circuits because of transmission and signaling considerations that will be discussed later. Each local or end central office also connects to its associated *toll switching office* or *toll center* serving that particular region. Calls between end offices (also designated Class 5) within the same area code region, for example 212, are normally routed over interoffice trunks and not through toll centers.

Toll centers (designated Class 4 offices) interconnect with Primary (Class 3), Sectional (Class 2), and Regional (Class 1) switching centers to form the Bell System's Direct-Distance-Dialing (DDD) Network within the United States, as illustrated by the diagram on page 7. Switching centers in the DDD Network are interconnected by underground coaxial cables, point-to-point microwave systems, and satellite radio systems. Carrier transmission techniques are employed to form the many thousands of voice and data channels provided by these communication facilities.

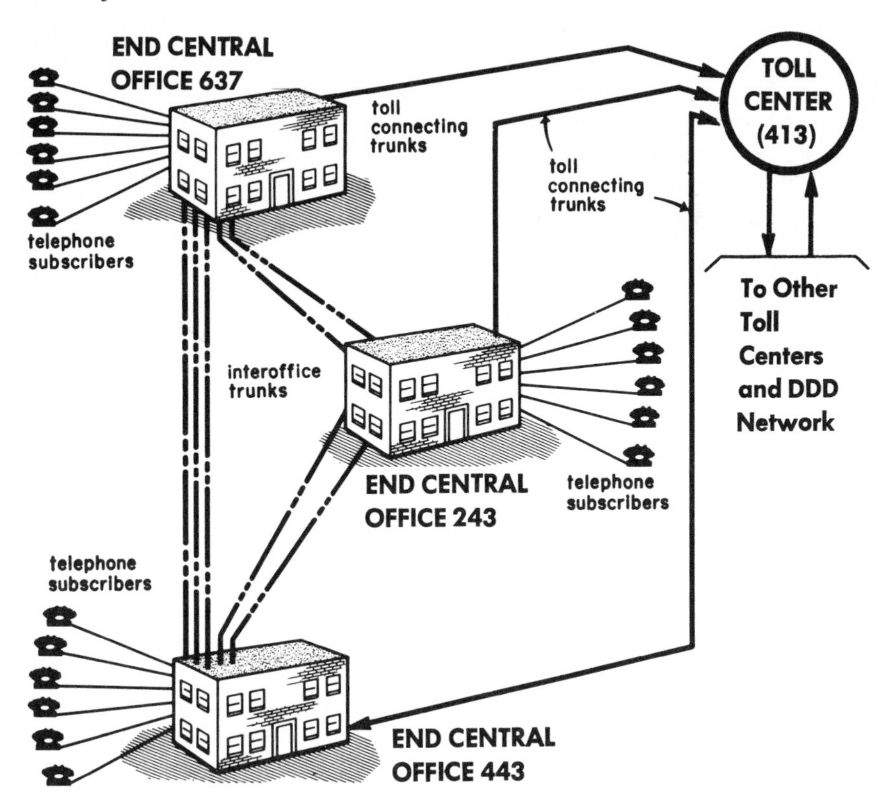

Elements of Telephone System

We should know what main items comprise a telephone system to better understand the important part played by carrier systems. In general, a telephone system can be divided into four main elements:

1. *Telephone instruments* in the homes or business offices of subscribers. They are commonly called *subscriber stations* or *subsets.*

Elements of Telephone System (cont.)

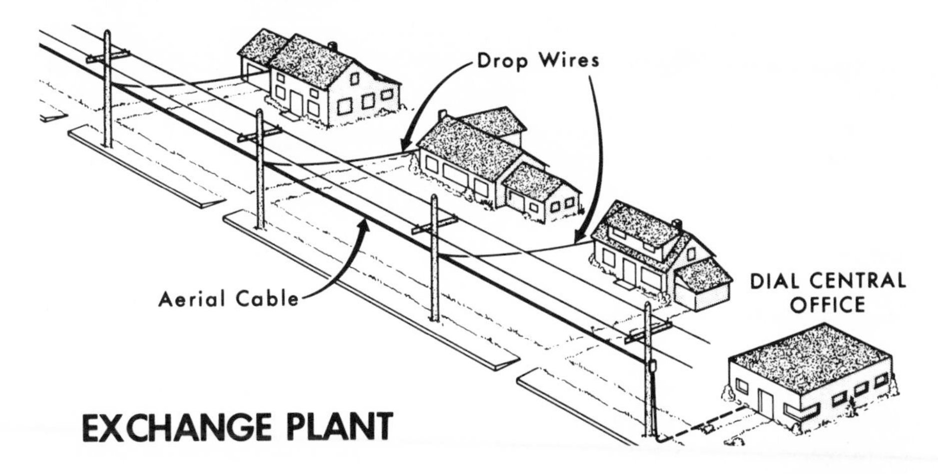

2. *Central offices* which switch the calls to the desired called party. These are dial operated, employing electromechanical and electronic switching equipment.
3. *Exchange plant,* the outside wire plant comprising the drop wires and aerial and underground cables which connect the telephone stations to the central office. The interoffice trunks are also included in this category.
4. *Toll plant,* the network of toll cables, coaxial cables, microwave radio systems, and satellite circuits which interconnect toll centers (Class 4 offices) with the Bell System's DDD Network.

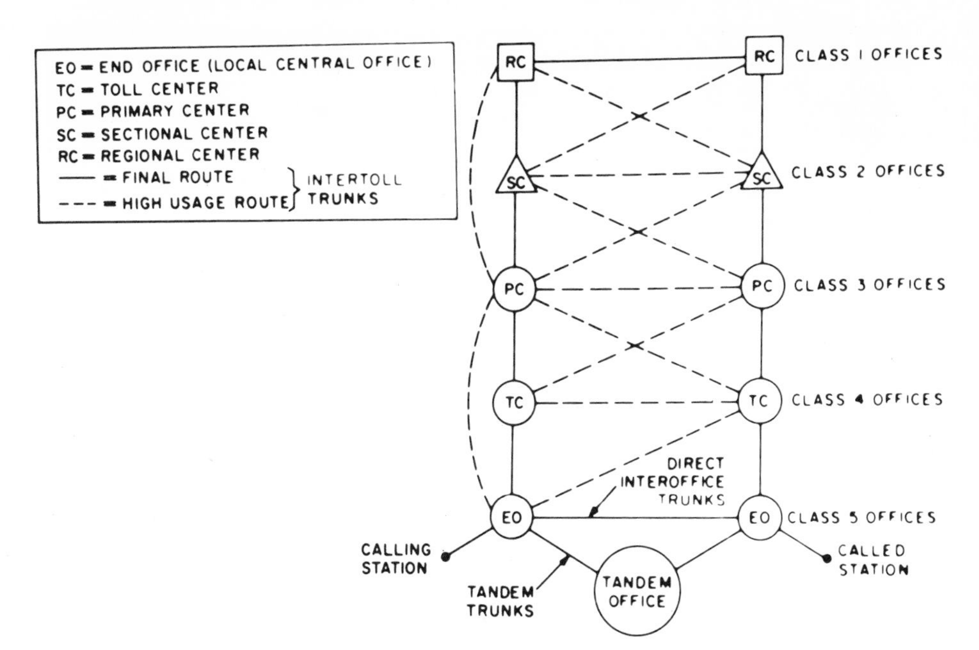

Need for Telephone Carrier Systems

A separate pair of wires, as previously mentioned, is required to connect each subscriber to his telephone exchange. Similarly, each interoffice trunk circuit which carries only one two-way conversation needs a pair of physical conductors. As the number of central offices increases, a considerably greater number of interconnecting trunk circuits will be required. The construction of outside toll plant facilities, such as microwave radio or aerial and underground cables, is very costly. In fact, if a physical pair of wires or a radio channel were limited to only a single two-way conversation, the rates for long distance calls would be very high. The development of *carrier systems* permits adding more conversations or voice channels to a physical wire circuit or radio channel.

The use of carrier or multiplex transmission over toll trunks, long distance wire, and radio circuits is the main consideration in this presentation of the principles of basic carrier telephony. Let us assume, for example, that the telephone cable between towns A and B has a capacity of only five toll circuits. With increasing telephone traffic between these towns, it becomes necessary to add two more circuits. Before the advent of carrier systems, it would have been necessary to replace the existing cable with a larger capacity type or to install an additional cable. This fact, along with the ever-increasing costs of material and labor, fostered the development of carrier transmission technology. Carrier systems can provide additional telephone channels on the same physical wire circuit or cable pair, as shown by the illustration.

EXAMPLE OF CARRIER APPLICATION TO TOLL CABLE CIRCUITS

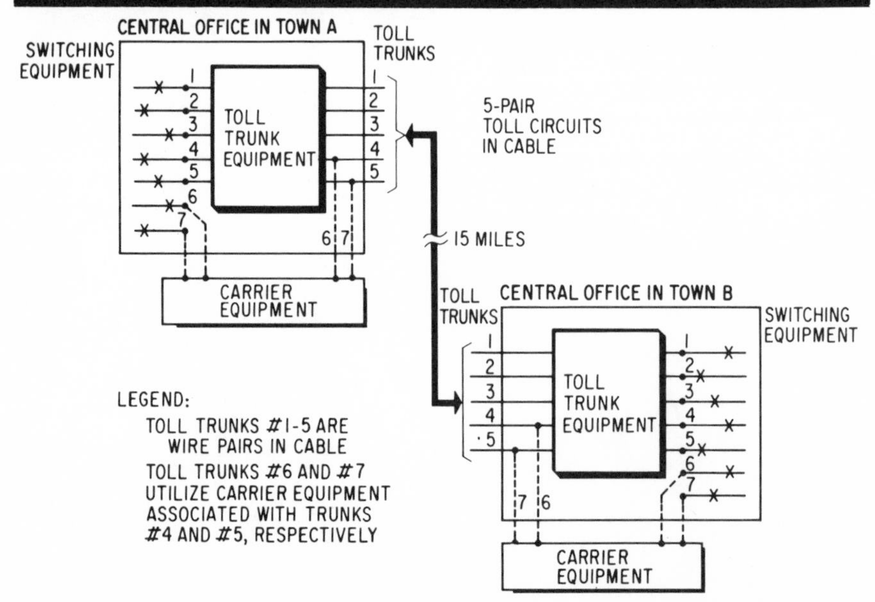

Need for Telephone Carrier Systems (cont.)

Modern carrier systems utilize solid-state devices which have great reliability. This factor, coupled with decreasing costs have made it more economical to provide carrier equipment rather than to install additional cables or other wire facilities for short-haul toll circuits and, in many instances, for rural subscriber lines.

Questions and Problems

1. What country is second to the United States in the number of telephones?
2. What percentage of the world's telephones is in the United States?
3. What percentage of total telephone revenues are derived from long distance or toll calls?
4. What Federal agency has jurisdiction over the interstate operations of telephone companies?
5. What office class designation is applied to the local central office? Toll office?
6. What is a wire center?
7. What is an interoffice trunk?
8. What is a toll center?
9. Describe the main elements that comprise a telephone system.
10. For what part of the telephone system was carrier equipment originally developed?

Characteristics of Speech

The telephone is used to send our voices in the form of electrical pulses over distances for communication purposes. We should, therefore, know something about the characteristics of speech to better understand the engineering requirements of a telephone system. Speech may be defined as a mechanical vibration of air within the frequency range to which the ear can respond. Pitch and loudness are the main characteristics of speech as applicable to the telephone.

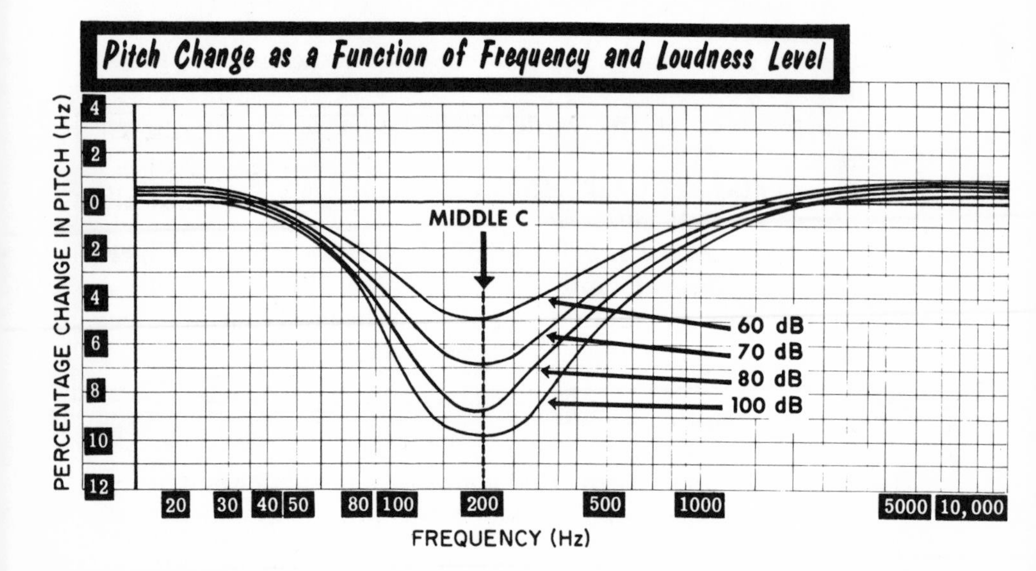

Pitch represents the frequency of vibration, and is expressed in cycles per second or *hertz* (abbreviated Hz). Most speech sounds are composed of complex waveforms having a number of components, the frequencies of which are harmonically related. These harmonics determine the *quality* of the speech. The fundamental frequency or single tone corresponding to these overtones or harmonics is the *pitch*. The fundamental pitch of the vocal chords varies with the individual but can be about 125 and 250 Hz for the normal male and female voices, respectively.

Loudness depends upon the amplitude of the sound wave. It may be expressed as the ratio of the energy in the actual sound wave to the energy contained in the weakest audible wave of the same frequency. Our speech can be generally subdivided into vowels, consonants, and diphthongs which convey the intelligence. A vowel sound is composed of the harmonics of the frequency of vibration, or pitch, of the speaker's vocal chords. Consonants have no fundamental frequency or pitch as they are formed by the tongue, lips, and palate without the help of the vocal chords. Each consonant is characterized by its own distribution of energy over the voice range.

Characteristics of Speech (cont.)

Diphthongs are a combination of two vowels in which the change from one vowel to another is accomplished without pause by changing the position of the mouth.

The various vowel and consonant sounds are identified by the listener's ears by the relative amounts of energy in the speech at the different frequencies. The distribution of this speech energy over the voice frequency range changes as different sounds are spoken. The following graph illustrates the speech frequency spectrum and the relative speech energy.

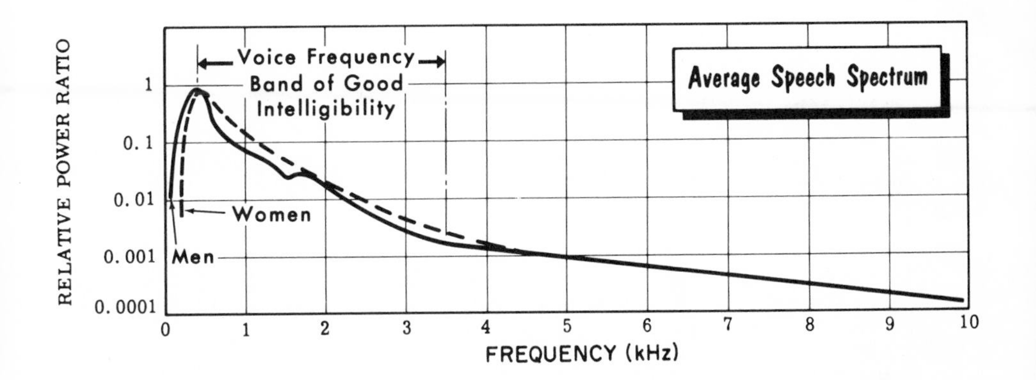

It will be observed that the speech frequencies cover a range of about 100 to 8,000 Hz and that most of the energy is concentrated at the lower frequencies around 250 to 500 Hz. Speech loudness or amplitude is determined mainly by these lower frequencies. In other words, most of the *power* or *energy* in our speech is contained in these lower voice frequencies.

The ear receives the sound waves so that its properties are also very important in telephone transmission. The curve below indicates the frequency and amplitude range of the normal ear. Note that the ear is most sensitive in the range of 1,000 to 3,000 Hz.

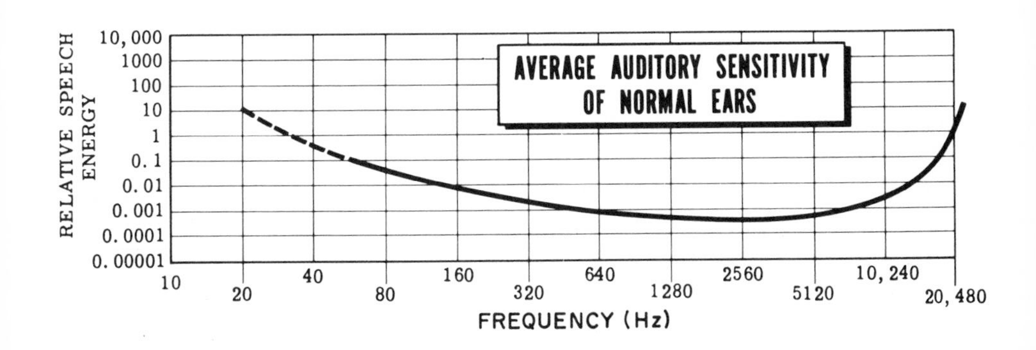

Voice Frequency Currents

We have learned that when we speak, the air is made to vibrate from a few to many hundreds or even thousands of times in a second. These vibrations or changes in air pressure can, by acting on the microphone of a telephone instrument, cause the generation of alternating electric currents. This conversion of *speech sounds* into a form of *electrical energy* and the subsequent change of this alternating or voice frequency current back to sound waves are the basis of telephony. It is essential that the reconstructed speech corresponds in frequency and waveform to the original sounds so that our conversation can be understood.

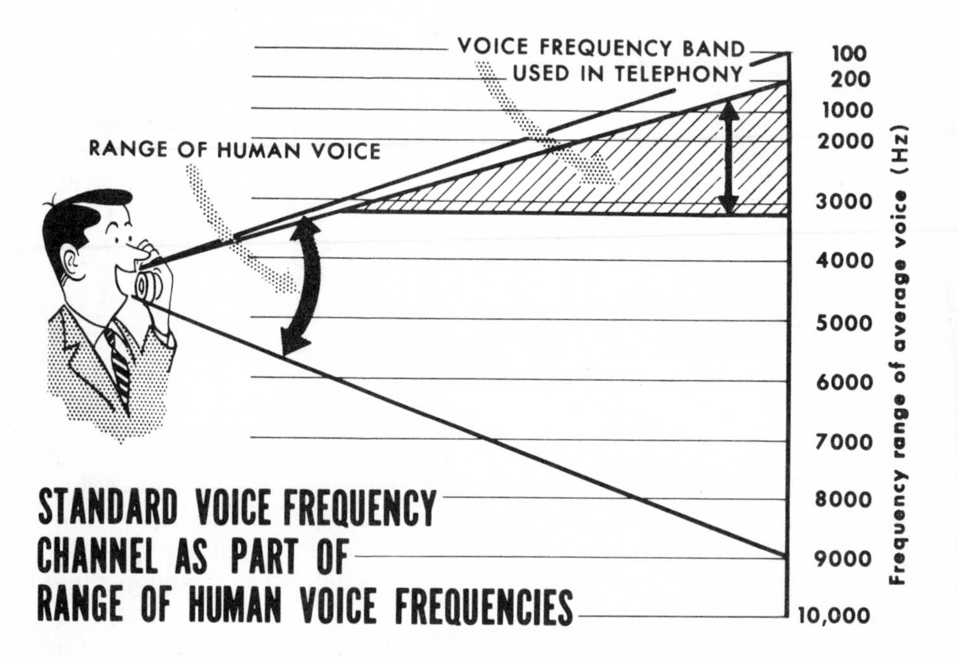

We also learned that ordinary speech usually covers a frequency range of about 100 to 8,000 Hz. It is not necessary, however, to transmit this entire band of voice frequencies for good speech intelligence. Satisfactory results are obtained if only the speech components in the range of approximately 200 to 3,400 Hz are transmitted over telephone circuits. In this connection, it is essential that the amplitude or loudness of the speech be maintained at a satisfactory level for good intelligence. If some of the frequency components of the voice currents are attenuated more than the others by the telephone line, the shape of the received speech wave heard by the listener will be changed. This may cause the received speech to become unintelligible. The frequency range of 200-3,400 Hz has currently become the *standard voice frequency channel* for transmission of speech over most wire and radio communication circuits.

Carrier Currents

As the name suggests, *carrier currents* are used to carry the speech intelligence in a telephone carrier system. They are a form of alternating current whose frequency is above that of the normal voice-frequency channel used in telephone communications. The selected frequency of this alternating or carrier current is generated by an oscillator in the transmitting terminal of the carrier equipment. The speech intelligence, in the form of voice currents, is superimposed on the carrier currents in a manner which we shall study later.

Carrier Currents "Carry" Speech Intelligence in a Telephone Carrier System

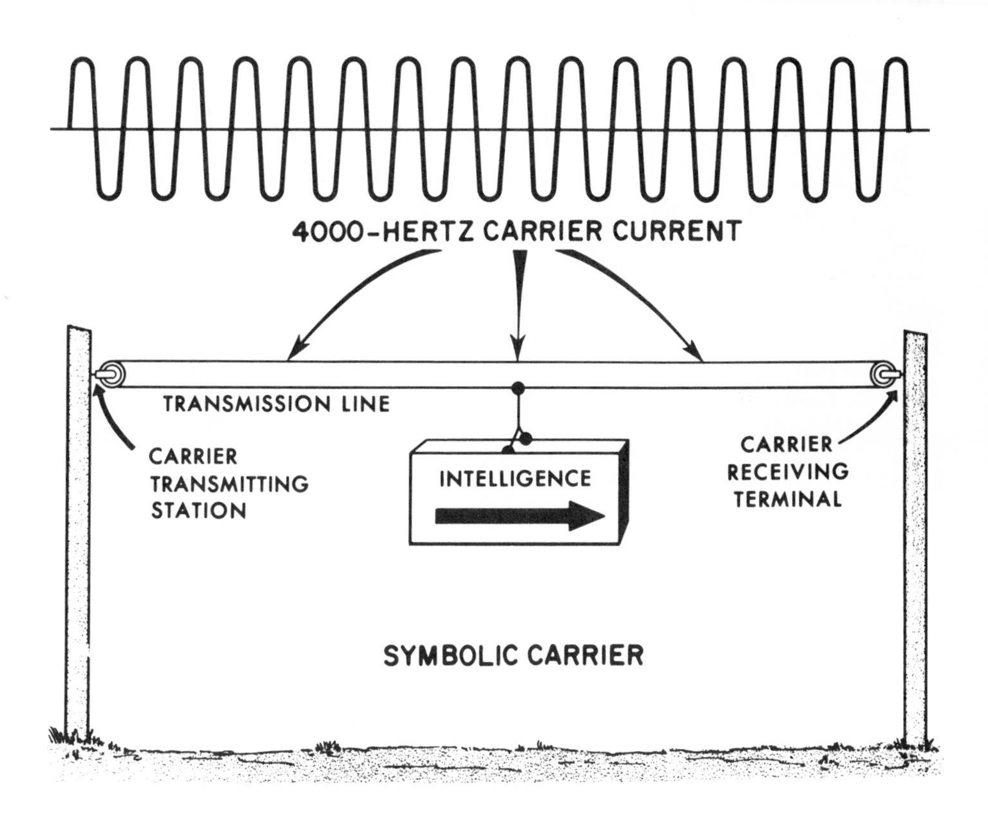

The resultant carrier frequency wave is transmitted along the wire lines to the carrier receiving terminal. Therefore, we may state that a carrier current is an alternating current (ac) of a certain determined frequency that can carry the voice current's intelligence that is superimposed on it.

Modulation

The superimposing of the voice currents on the generated carrier frequency is called *modulation*. This process or mixing action is made possible by a device called a *modulator*. One of the more important effects of the modulation process in a carrier system is the shifting or reallocation of the original voice frequency band.

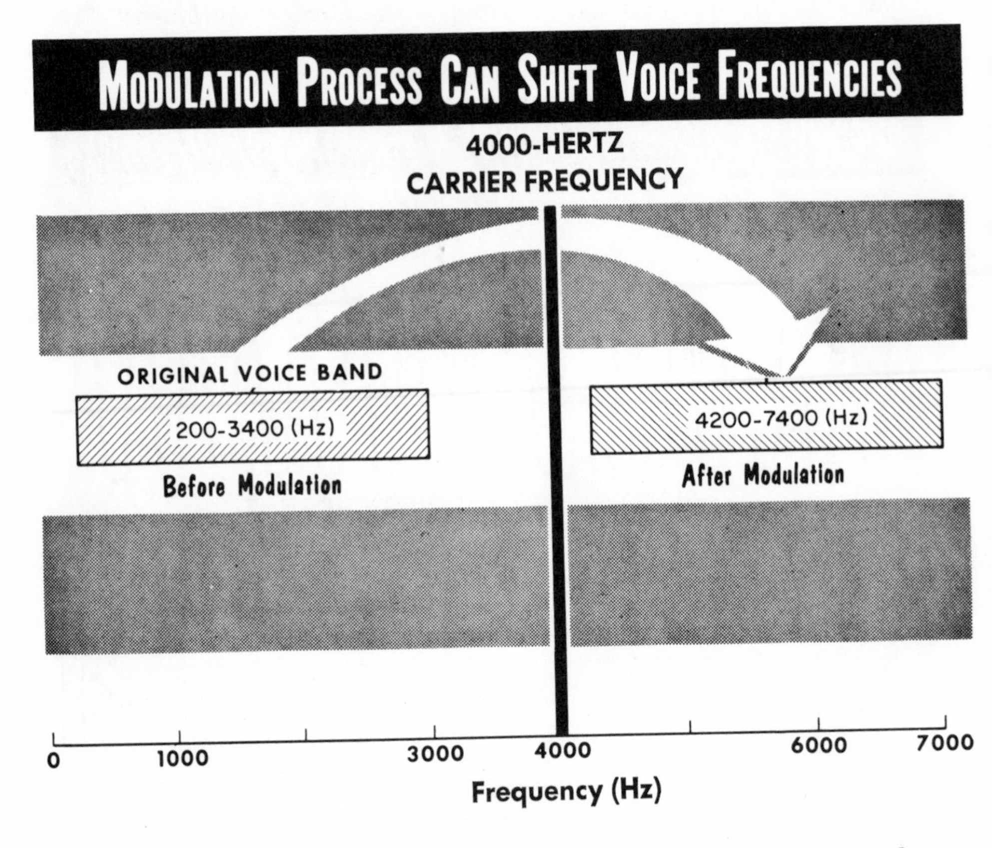

You will recall in the discussion on voice frequency currents that the frequency range of about 200 to 3,400 Hz had been standardized as the bandwidth of a telephone voice channel. The modulation process, as will be described later, can cause this band of voice frequencies to be shifted, by a generated carrier frequency, to another part of the frequency spectrum. For example, the 200-3,400-Hz speech band could be changed to occupy the 4,200-7,400-Hz range by a suitable carrier frequency. This will permit two separate telephone conversations to use the same pair of telephone wires without mutual interference. One conversation would occupy the normal 200-3,4000-Hz voice frequency band and the other one could utilize the 4,200-7,400-Hz range as a result of the modulation process.

Demodulation

Generally speaking, *demodulation* is the reverse of *modulation.* The received carrier signal on which the speech intelligence had been previously superimposed is combined with the same carrier frequency as was originally used in the modulation process. This carrier signal is reinserted in a device called the *demodulator.* In a manner similar to that discussed in the section on modulation, the original voice frequencies are *separated* from the carrier signal.

DEMODULATION SEPARATES VOICE INTELLIGENCE FROM CARRIER

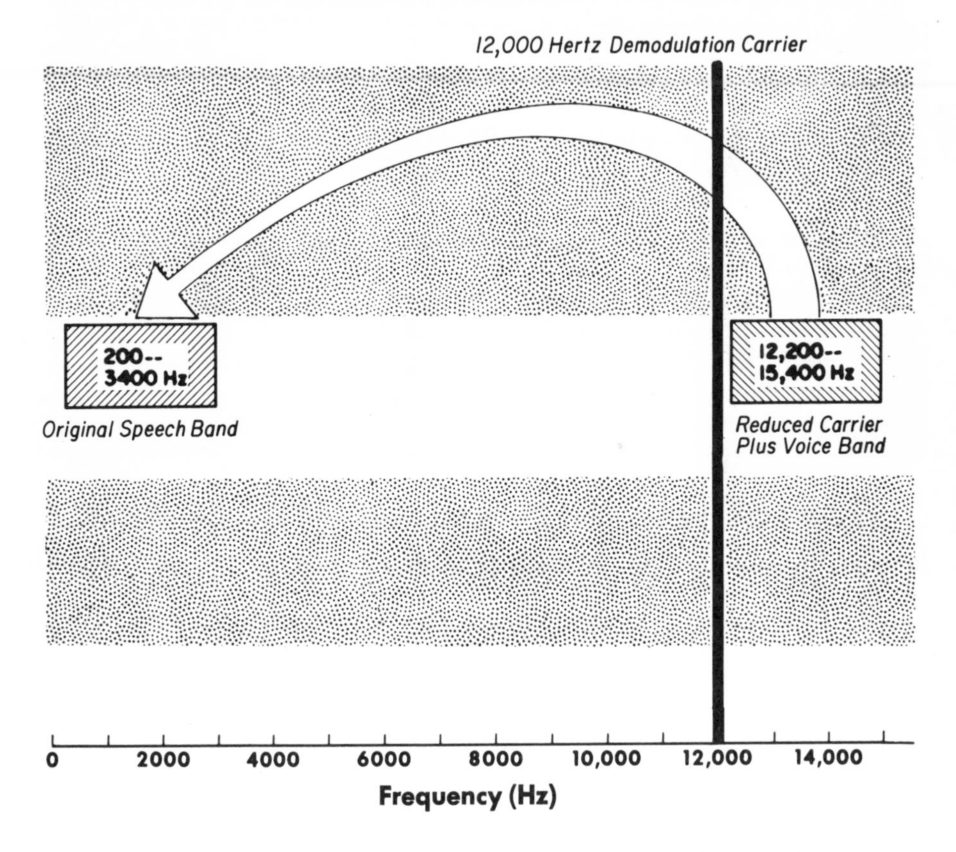

If we assume that the received carrier signal with its superimposed modulation has a range of 12,200 to 15,400 Hz, it will be necessary to re-insert a carrier frequency of 12,000 Hz to recover the original 200-3,400 Hz voice frequency band. The process of separating the voice intelligence from the carrier, as described above, is defined as *demodulation.*

Sidebands

The resultant output of the modulation process contains three basic elements: the inserted carrier frequency, and the two bands of new superimposed frequencies called *sidebands.* The *sum* of the carrier frequency and the speech frequency range which modulated the carrier is the *upper sideband.* The carrier frequency *minus* the voice band is called the *lower sideband.* For example, if the carrier oscillator's frequency is 12,000 Hz and it is modulated by a single tone of 500 Hz, the modulator output signal will contain the following frequencies:

a. 12,000 + 500 = 12,500 Hz.......................... Upper Sideband
b. 12,000 Hz.................................. Carrier Frequency
c. 12,000 − 500 = 11,500 Hz.......................... Lower Sideband

Note that both the lower and upper sidebands contain the original modulating 500-Hz tone. The carrier is the single frequency that was used for translation purposes in the modulator. Note also that it is only necessary to transmit *one* of the sidebands to convey the 500-Hz tone or, for that matter, any speech intelligence. This method is employed in many of the modern carrier systems in the telephone plant. The carrier and one sideband are suppressed or filtered out, and the other sideband is transmitted over the wire line.

Sidebands also are generated in a similar manner in the demodulation process. Let us assume that an incoming carrier signal has a range of 24,200 to 27,400 Hz. By inserting a 24,000-Hz carrier in the demodulator, the resultant output will comprise:

Upper Sideband: 48,200 to 51,400 Hz

Lower Sideband: 200 to 3,400 Hz (the original speech band)

The recovered lower sideband which contains the 200-3,400-Hz voice band is utilized for transmission to the toll center and to the other telephone station; the upper sideband frequencies are filtered out. The 24,000-Hz carrier is usually suppressed within the demodulator. In summation, when two different frequencies are combined in a modulator or demodulator, the sum and difference of the two original frequencies appear in the output signal. The sum is termed the upper sideband; the difference is called the lower sideband.

RESULTANT OUTPUT OF MODULATING CARRIER WITH 500-HERTZ TONE

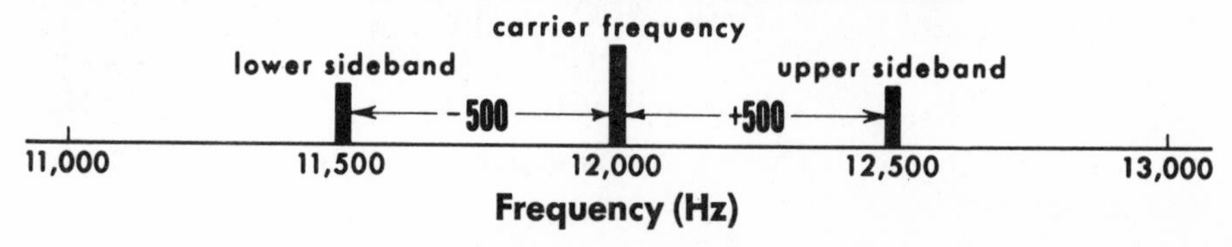

Filters

Just as a Chemical Filter can separate solid matter from liquids...

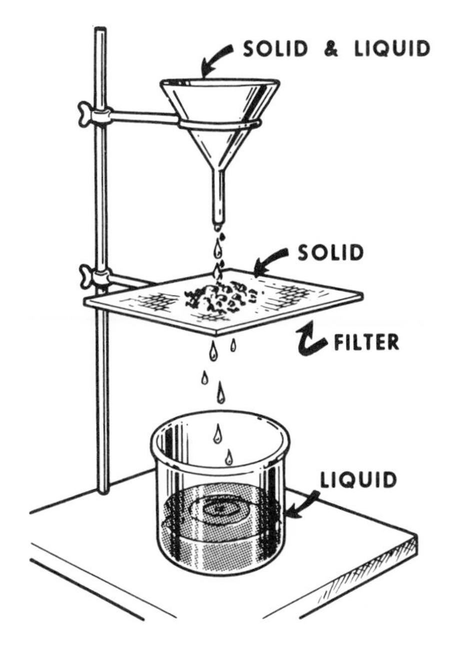

an Electrical Filter can suppress or reject certain frequencies while passing others.

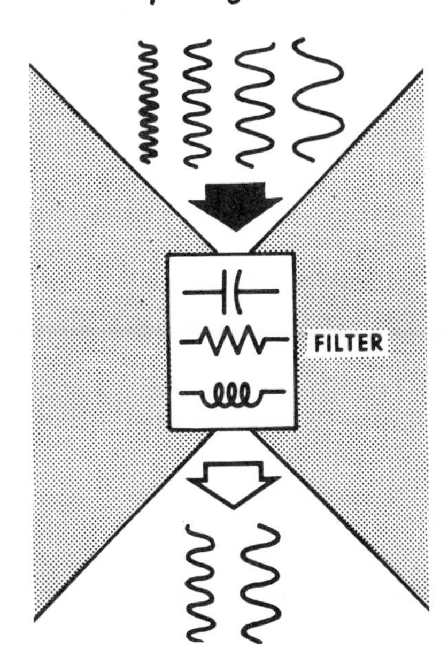

We are probably familiar from our high school chemistry days with the use of filter paper to separate solid matter from a liquid. The same principle can be applied to electrical circuits. It often is necessary to eliminate or suppress certain frequencies that are not needed or that will hinder effective transmission of voice currents. This is particularly true in the modulation process where several different frequency components may be present and it is desired that only one group be transmitted. To accomplish this, a circuit is designed that will give a much higher loss to the unwanted frequencies than to the desired ones. This circuit and its components comprise an electrical filter, more commonly called a *filter*.

The loss introduced by filters varies with frequency. Filters normally consist of circuit components which have impedances that vary with frequency. This usually requires the use of combinations of inductance and capacitance values (coils and capacitors) whose respective reactances will fulfill the filter design requirements.

Four general types of filters are usually employed in carrier systems of the analog type. They are commonly designated: *low-pass* filter; *high-pass* filter; *bandpass* filter; *band-rejection* or *band-stop* filter. The design and use of these various types of filters will be described and illustrated later.

Repeaters

It often is necessary to amplify the voice signals because of the losses incurred in transmission over wire lines. The electronic device used for this may be compared to the audio amplifier of a radio broadcast receiver or a record player. While it is possible to use one amplifier in a circuit arrangement to amplify two-way telephone conversations, two separate amplifiers normally are used, one for each direction of conversation. These amplifiers with their associated hybrid coils, balancing networks, and other apparatus comprise a *repeater.*

There are several types of voice-frequency repeaters that have been employed in the telephone plant for many years. The Western Electric V3 and 24V4 and the GTE Automatic Electric E-6 are some of the more common types. All of these repeaters are designed to utilize hybrid coil arrangements. With the exception of the E-6 (negative impedance) type, all use two separate amplifiers to serve both directions of conversation. The diagrams illustrate in simplified form the application of these repeaters to a toll circuit. The 44V4 repeater is employed for four-wire operation, that is, where separate pairs of wires or channels are used to carry each direction of conversation. This method is used for most cable carrier systems. The terminal type repeater is often utilized to connect such four-wire circuits to the usual two-wire central office trunk. The V3 and E-6 repeaters are designed for two-wire operations as in the usual telephone circuit. The amplifiers in these repeaters are transistorized.

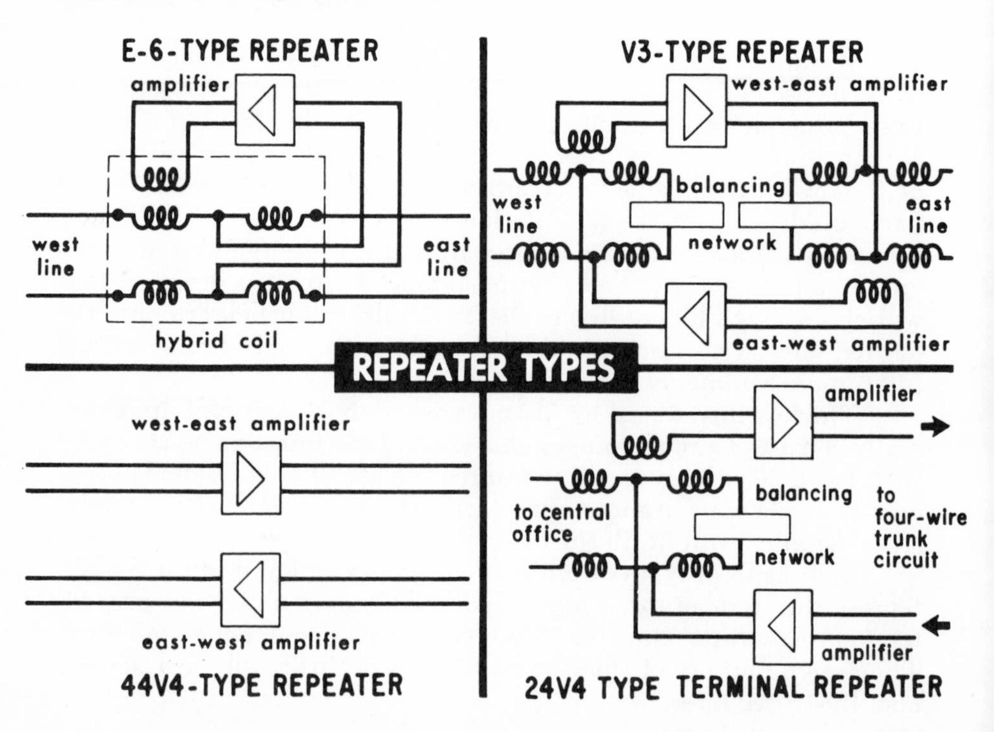

Repeating Coils

Typical Repeating-Coil Circuits

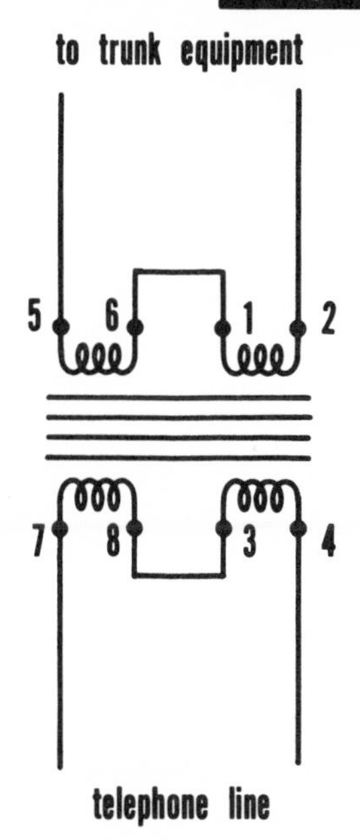

COILS 1-2, 3-4, 5-6, AND 7-8 ARE ALL EQUAL WINDINGS.

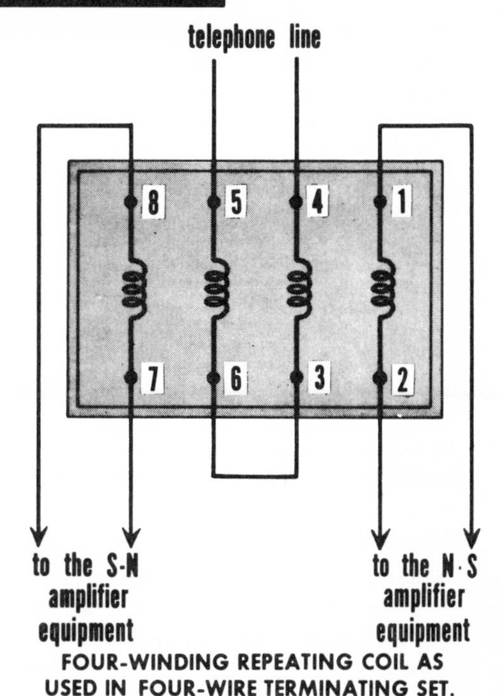

FOUR-WINDING REPEATING COIL AS USED IN FOUR-WIRE TERMINATING SET.

Most of us are familiar with power transformers and their use. For example, the transformer in the power supply of our television set is used to step up the normal 120 volts for subsequent rectification. The bell transformer which rings our door bell steps down the line voltage from 120 volts to about 6-15 volts. These transformers change the voltage and current values as required for their particular applications. In the telephone plant, a transformer is often used to repeat the energy from one circuit to another without changing the voltage or current values. This transformer arrangement is called a *repeating coil.*

As a general rule, repeating coils consist of two or more windings, each magnetically coupled to all the other windings. The strength of a magnetic field, as in a transformer, depends on the product of the number of turns and the amount of current in the coil winding. Therefore, in a repeating coil of two windings, the two magnetic fields must be equal in magnitude to repeat the voice signals from one circuit to the other without voltage or current changes. This is accomplished by using windings of the same number of turns to give a 1 to 1 ratio of transformation. The diagrams show some common repeating coil circuit arrangements.

Hybrid Coils

In the definition of repeaters, we learned that two amplifiers normally are required; one for each direction of conversation. If, however, these two amplifiers were connected into the telephone line as shown, the circuit would not work because of the feedback of the energy. The amplifiers would howl or *sing,* which is a sign that they were unstable in operation.

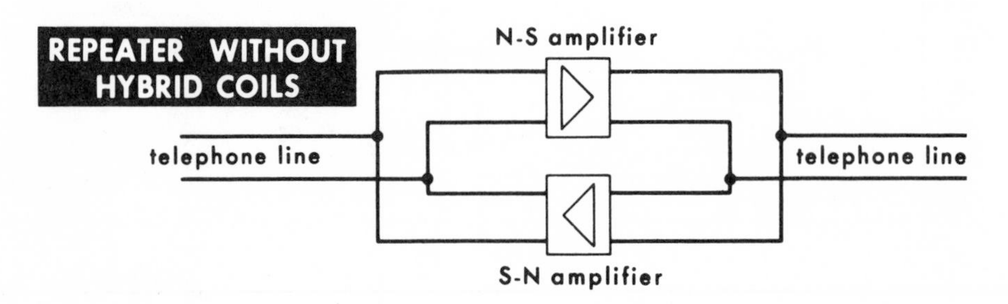

To enable the two amplifiers to be used successfully for two-way transmission over a pair of wires, it is necessary to add certain equipment. The most important requirement is a transformer of hybrid coils or equivalent devices. Four such common arrangements are called: (1) hybrid coil; (2) repeating coil hybrid, using repeating coils; (3) four-wire terminating set, which employs a four-winding coil arrangement; and (4) resistance hybrid, which uses resistors in a type of Wheatstone bridge circuit instead of transformers.

The hybrid coil is used extensively in repeaters and carrier terminal equipment. It is a transformer that has three pairs of balanced coils mounted on a common permalloy type core. These coils are generally termed primary, secondary, and tertiary windings. They are connected as shown. By inserting a hybrid coil in the circuit of each amplifier, we obtain a repeater that can amplify two-way conversations without feedback or *singing* effects.

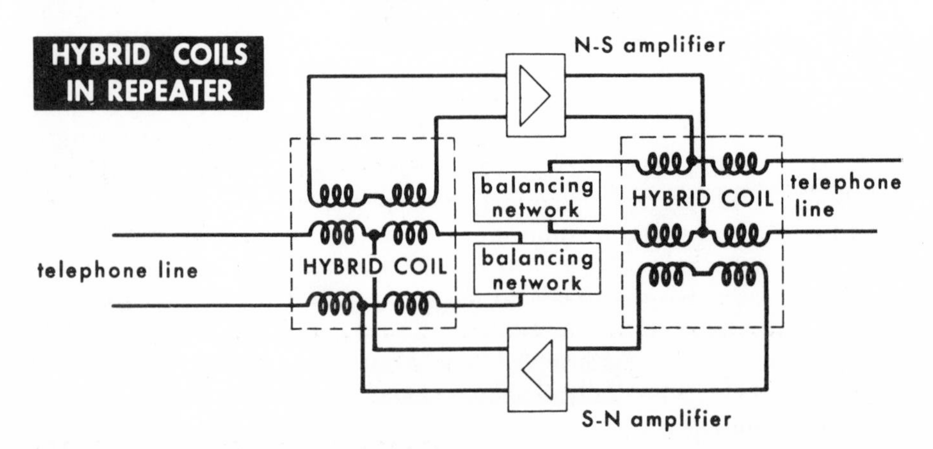

Equalizers

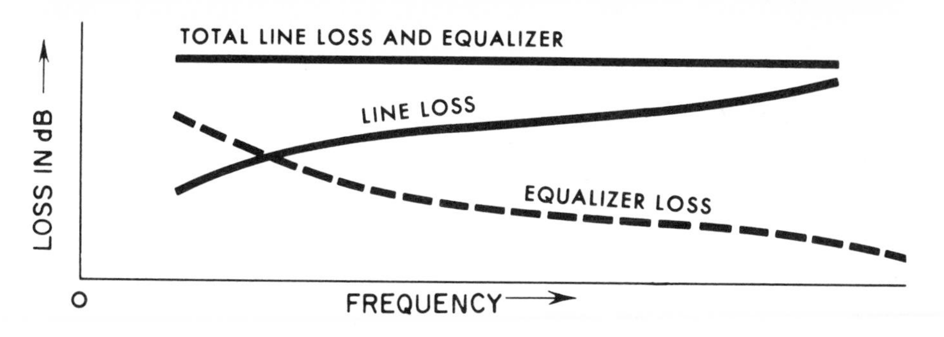

The losses in transmission lines, as we shall see later, are not the same at all frequencies. High frequencies are attenuated to a greater degree than low frequencies and this causes distortion of the transmitted signals. It is necessary, therefore, to use a network or combination of resistive, capacitive, and inductive components to correct the unequal attenutation in long telephone lines. The apparatus used for this purpose is called an *attenuation equalizer* or just *equalizer.* The purpose of the *equalizer* is to add frequency selective attenuation into the telephone line circuit so that the total loss or attenuation will be about the same for all frequencies transmitted. Telephone lines used to carry radio broadcast programs from the studio to the transmitting station, for example, are usually *equalized.* That is, an equalizer is added to the circuit so that all frequencies in the required range will be amplified equally.

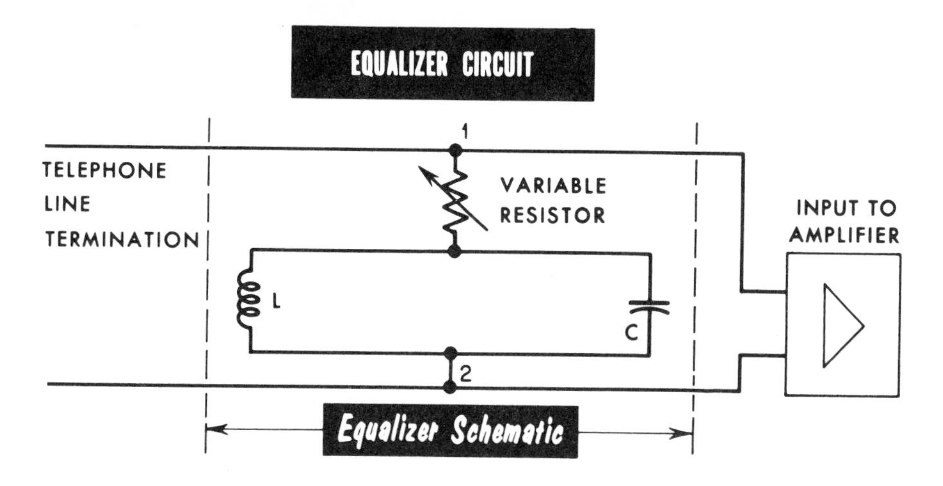

Attenuation Pads

In the operation of telephone carrier equipment, it is desirable to transmit the maximum amount of power to the telephone line at all frequencies used in the carrier system. Since the losses in a telephone line vary with frequency, it is the usual practice to attenuate the different frequency components of the signal equally. In this manner, all frequencies transmitted to the telephone line will be of about equal power. This is best accomplished by a *resistance pad,* or *pad* as it is commonly termed.

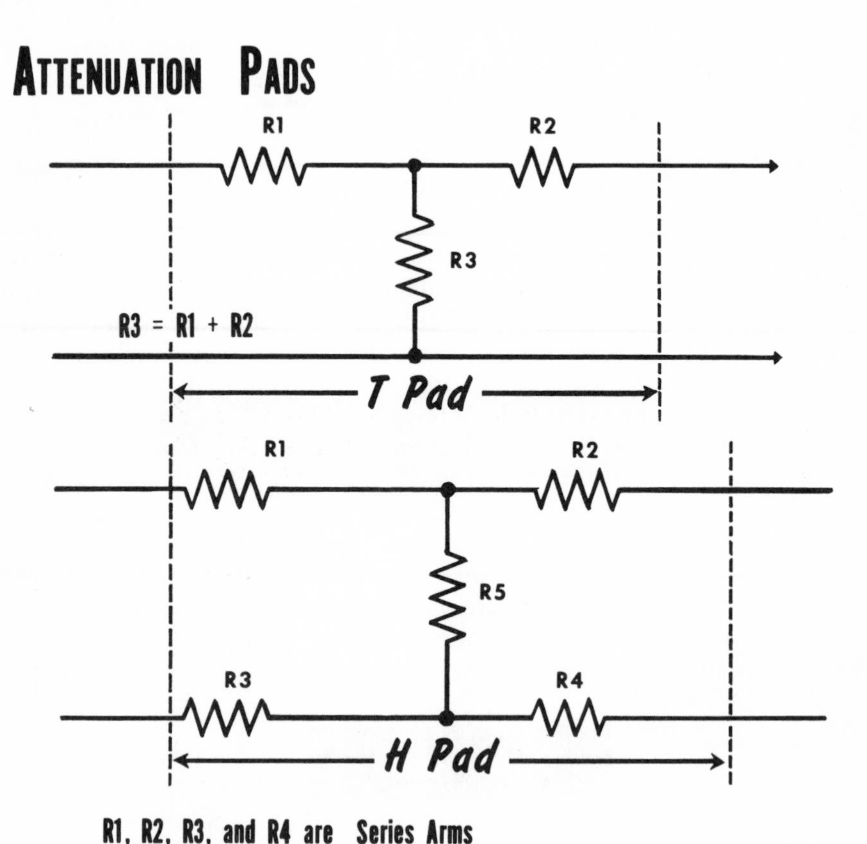

As the name indicates, the pad is a network of resistors of known values selected for the particular needs of the carrier equipment. Several common types of pads generally employed in telephone equipment are designated T, H, Pi, and the square pad. The application of these pads will be described later.

Balancing Network

The improvement in transmission afforded by the use of repeaters depends to a great extent on the impedance match or balance of the telephone line and the hybrid coils in the repeater equipment. To provide the proper balance, a network of resistive, capacitive, and sometimes inductive components is connected to the hybrid coils. This circuit arrangement is designated a *balancing network.*

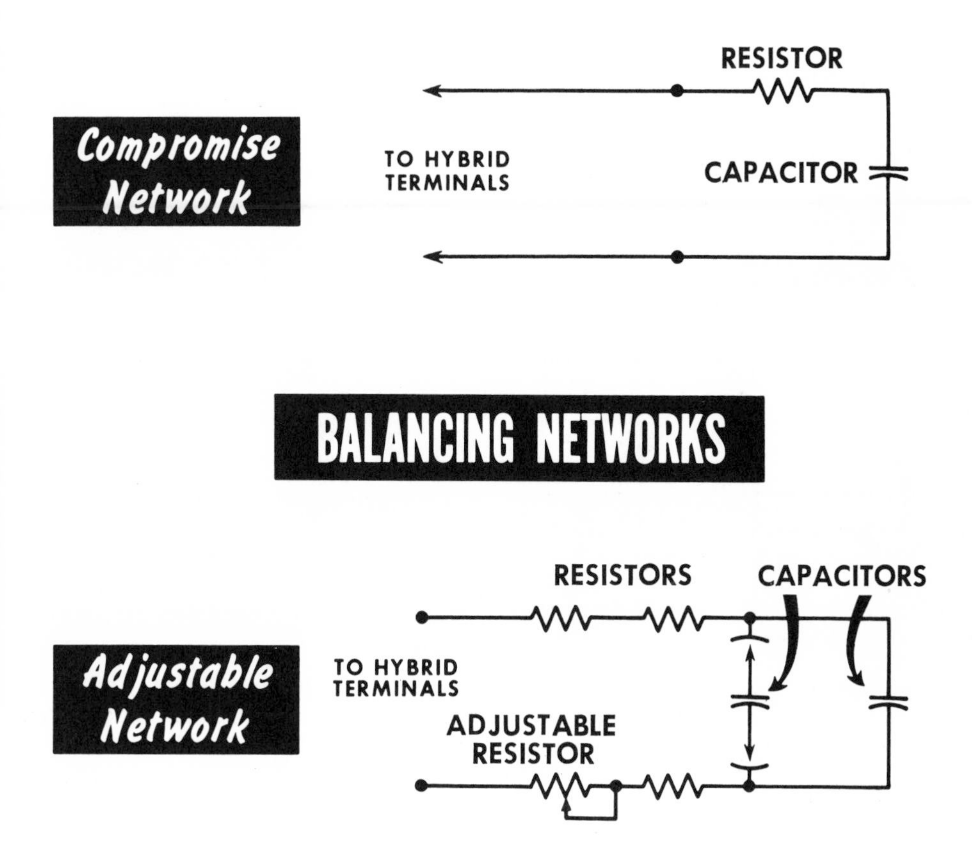

Since the telephone line impedance varies with frequency, the balancing network must be able to cover the frequency range of the repeater. Some basic types of balancing networks are the compromise, precision, and adjustable networks.

Semiconductors

Semiconductors or solid-state devices, such as diodes, transistors, and zener diodes, have the property of offering low resistance to current flow in one direction and very high resistance in the reverse direction. They have replaced vacuum tubes in carrier, wire transmission, power supplies, and other telephone equipment. These solid state devices also have replaced vacuum tubes in mobile radio-telephone units and microwave radio apparatus. The small size of semiconductors (which contain germanium, silicon, and other crystal elements) and their minute power requirements make them particularly advantageous for employment in carrier, repeaters, radio relay, and other telephone facilities. The new electronic switching systems, which are replacing present electromechanical dial central offices, use solid state devices in large quantities. Likewise, the new *Touch-Tone*

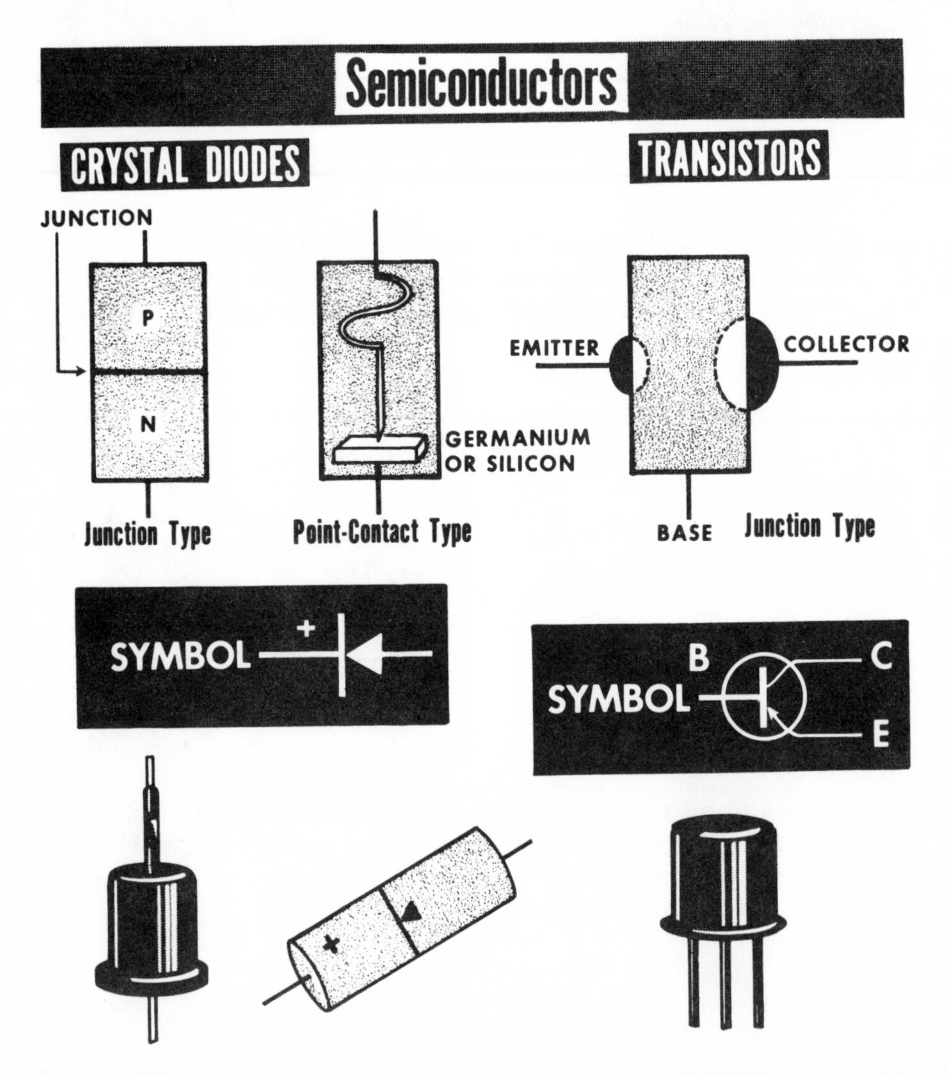

Semiconductors (cont.)

telephone instrument, which is replacing the rotary dial, employs semiconductors for generating audio signaling tones.

Recent developments have made it feasible to combine a number of semiconductors and other circuit elements into an "integrated circuit." These microminiature devices, which are extremely small (about the size of the proverbial *head of a pin*), have many applications in the telephone plant, especially in electronic switching systems.

Questions and Problems

1. What speech characteristics are most important to telephone transmissions?
2. In what voice-frequency range is most speech energy concentrated?
3. What range of voice frequencies is usually considered the standard for telephone transmission?
4. What are carrier currents?
5. Define modulation.
6. What is a demodulator?
7. What basic elements result from the modulation process?
8. What electrical characteristics are possessed by a filter?
9. What is a repeating coil?
10. What electrical property is possessed by a solid-state diode?

Characteristics of Long Telephone Lines

A transmission line, generally speaking, is a means of transferring electrical energy or power from one point to another. The electric power company, for example, employs high-voltage transmission lines usually installed on high steel towers to carry the electric power from the generating station to the substations in the city. Lower voltage transmission lines are then used to distribute the electric power to the factories and homes. In telephone communications, the magnitude of the signal power is extremely small—in the order of milliwatts rather than the hundreds or thousands of kilowatts transmitted by the electric power distribution system. The losses in the telephone transmission line, therefore, are very important to the efficient operation of the telephone system.

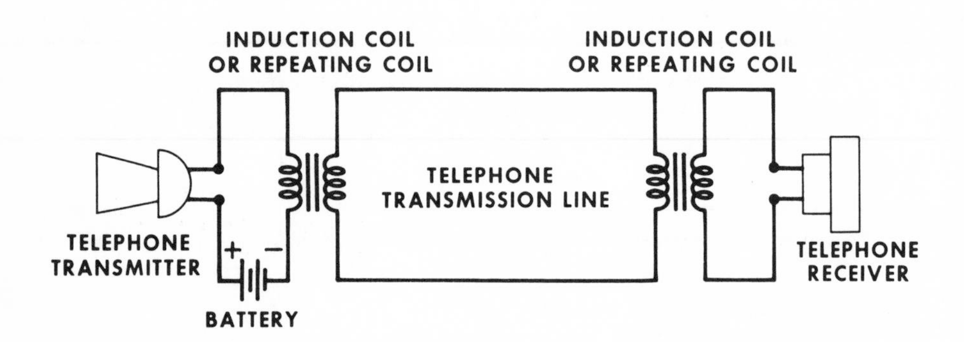

ESSENTIAL PARTS OF TELEPHONE SYSTEM

Just as in the case of electric power distribution, a telephone communication system has three essential parts:

1. A telephone transmitter to generate electricity from the sound power of the speaker's voice.
2. The transmission line or pair of wires to carry this minute electric power. (The telephone transmission line, like the electric power line, must contain two conductors although one can be the ground or earth itself.)
3. A device such as the telephone receiver to convert this electric energy back to sound.

Characteristics of Long Telephone Lines (cont.)

So that we may better understand the electrical properties of a telephone line, let us consider a lamp load on an ac generator. When a load like a 100-watt lamp (lamp 1) is connected directly to a small 120-volt ac generator, it will light to full brilliancy. However, if we should now add about 300 feet (91.4 meters) of small-gage wire to connect a lamp (lamp 2) to the power source, it will not light so brightly. The power consumed by lamp 2 is less than that supplied by the ac generator. Therefore, power must be dissipated in the connecting wires or transmission line.

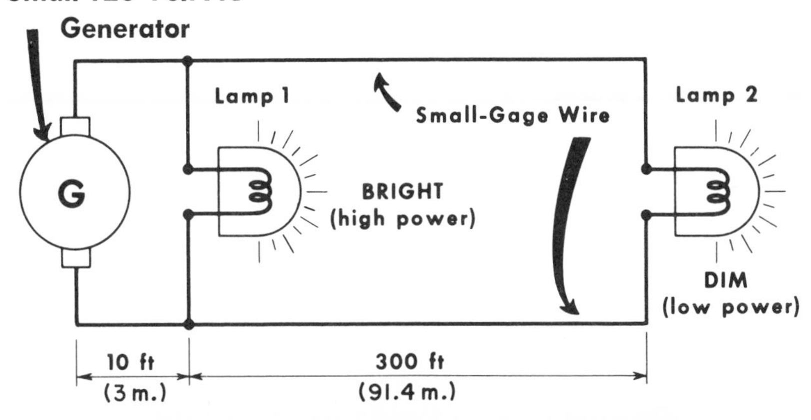

Dissipation of Power in Transmission Line

A similar action takes place in a long telephone line. This power loss does not take place at any one point in the line. It occurs in equal steps or increments along the length of the line. We can say, therefore, that the electrical characteristics that cause this power loss are uniformly distributed over the entire length of a long telephone line.

There are four main characteristics or electrical properties of a telephone line. These are usually termed *distributed constants per unit length* and are known as follows: (We shall discuss them in more detail later.)

	Symbol
1. Series Resistance	R
2. Series Inductance	L
3. Shunt Capacitance	C
4. Shunt Conductance or Leakage	G

Series Resistance

The series resistance to direct current (dc) of a telephone line can be found in the same way as determining the resistance of a length of wire by Ohm's law. Its dc resistance per a unit length can be measured and this value multiplied by the total length of the line. For example, if a 5-mile (3.125 km) long telephone line has a dc resistance of 10 ohms per 1,000 feet (304.8m) of wire, its total resistance will be:

$$26{,}400 \text{ (feet)} \times 10 \text{ (ohms)}/1{,}000 \text{ (feet)} = 264 \text{ ohms for one conductor}$$

The total resistance for the two-wire line length of 5 miles is:

$$264 \text{ ohms} \times 2 = 528 \text{ ohms}$$

It is usual in telephone terminology to express the resistance of a telephone line in *ohms per loop mile.* Thus, in the above example, the telephone line would have a dc resistance of about 105 ohms per loop mile for the two conductors.

Telephone lines carry voice frequency and carrier currents (forms of alternating currents) in addition to direct currents. When voice frequency or carrier currents are sent over the line, its distributed resistance depends on the frequency of these currents as well as the size of the wire. This variation of resistance with frequency is the result of the *skin effect* phenomenon. That is, the higher the frequency of an alternating current, the more the current tends to travel along the outer surface of the conductor rather than inside the wire. The series resistance, or R, of the telephone line will be greater than its dc resistance to the voice frequency currents.

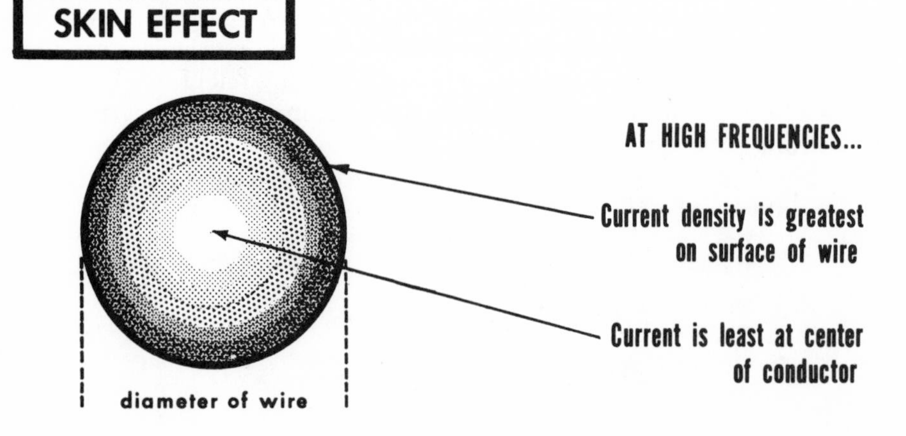

Series Inductance

It will be recalled from our study of ac theory that when an alternating current flows through a coil of wire, there is a counter-electromotive force or voltage induced in the coil which opposes the ac voltage. This property of the coil or circuit is called its *self-inductance* or *inductance.* In a long telephone line through which a changing current is flowing, a countervoltage is induced along the line. This indicates that the line has series inductance distributed along its entire length. The amount of this series inductance, or L, can be determined by the size or gage of the wires and their separation in the cable pair. The inductance will increase as the distance between the two wires increases; it decreases as the diameter of the wires increases. Series inductance is usually expressed in millihenries (mH) per loop mile.

The series inductance of a long telephone line opposes the voice frequency and carrier currents traveling along the line because of its *inductive reactance.* This inductive reactance, X_L, depends upon the frequency of the alternating current.

The formula below shows that the inductive reactance, or opposing effect of the induced voltage due to series inductance, increases with frequency.

Formula for Inductive Reactance

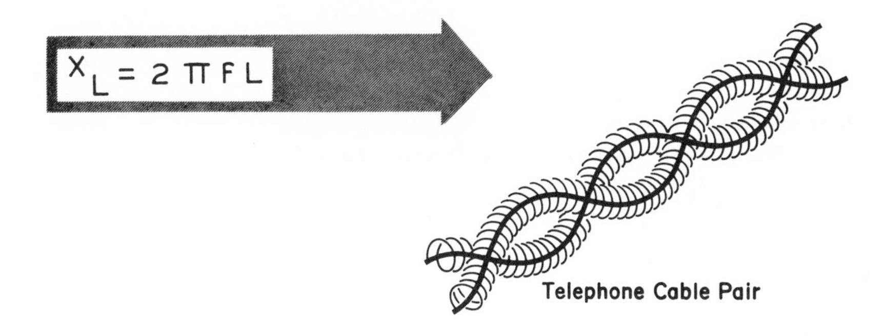

WHERE:

X_L ---- INDUCTIVE REACTANCE, EXPRESSED IN OHMS

2π ---- CONSTANT, USUALLY EXPRESSED AS 6.28

F ------ FREQUENCY, IN HERTZ (Hz)

L ------ SERIES INDUCTANCE, IN HENRIES

Shunt Capacitance

FORMULA FOR CAPACITIVE REACTANCE

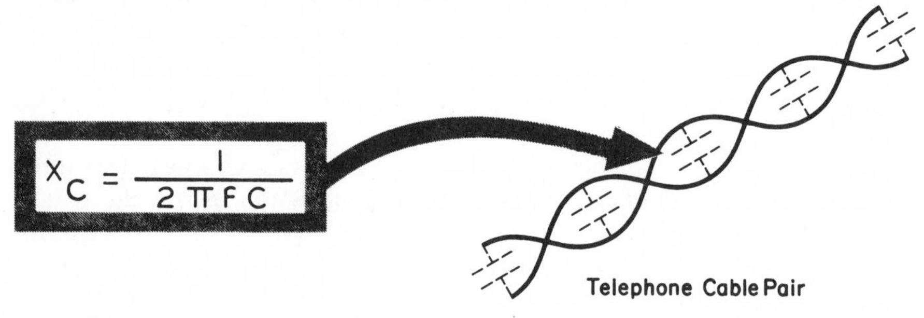

WHERE:

X_C ------ CAPACITIVE REACTANCE, EXPRESSED IN OHMS

2π ------ CONSTANT, USUALLY EXPRESSED AS 6.28

F ------- FREQUENCY, IN HERTZ (Hz)

C ------- CAPACITANCE, IN FARADS

Many years ago, open-wire telephone lines consisted of two-wire conductors separated by several inches of air and supported on insulators on the pole cross-arms. This can be considered to comprise a capacitor with air as the dielectric. As in the case of the ordinary capacitor, the capacitance is increased when the diameter of the wires is made larger. The capacitance also increases as the distance between the two conductors is made smaller. In other words, the shunt capacitance of the telephone line increases with the size of the wire and as the distance between the centers of the wires decreases. In addition, the capacitance is greater at the glass insulators on pole cross-arms because of the change in the dielectric constant from air to glass. In a telephone cable, the capacitance is much greater than on open-wire lines because of the very close spacing of the conductors and the use of insulating material as the dielectric instead of air.

This distributed capacitance appearing between the adjacent wires of the transmission line or cable pair is termed the *shunt capacitance.* The symbol is C and the capacitance is usually expressed in *microfarads per loop mile.* This distributed capacitance causes a capacitive reactance (Xc) to develop across the cable pair. As a result, some of the voice frequency currents are shunted across the line and less of the energy will reach the end of the line. The capacitive reactance decreases as the frequency increases so that this shunting effect on the voice and carrier currents becomes greater with increased frequency.

Leakage or Shunt Conductance

In discussing the effects of shunt capacitance, we learned of the importance of the dielectric constant. The air between the wires on a pole line and the glass insulators that separate the wires on the supporting cross-arms may be considered as dielectrics. Paper, plastic, and other insulating materials used in cables are not perfect insulators. As a result, a leakage current flows between the two conductors of the telephone line at the insulators on the cross-arms and between the conductors in a cable pair. During heavy rains, snow, sleet, or humid weather conditions, there is also leakage through the air between open-wire lines. In cables, temperature and humidity variations cause changes in the amount of leakage between the conductors of a cable pair.

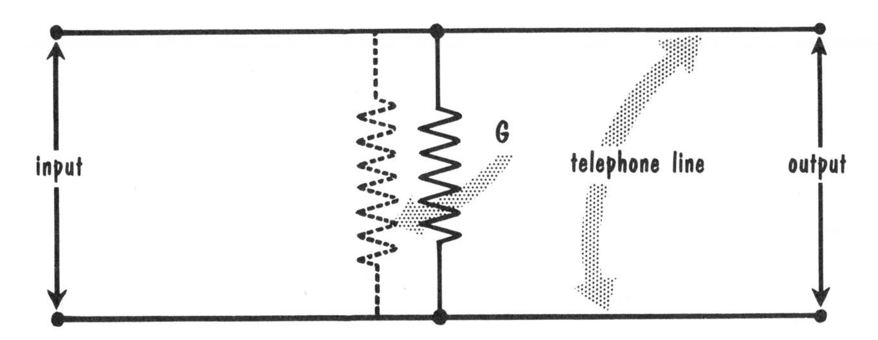

DISTRIBUTED SHUNT CONDUCTANCE OR LEAKAGE, G, IN A LONG TELEPHONE LINE

In general, leakage will take place at every point along the transmission line. It is usual to term it *shunt conductance* or *leakage* with the symbol G. This leakage acts as a shunt across the telephone line to the voice currents in a similar manner as that caused by shunt resistance. It is, therefore, an additional loss to the transmission of voice or carrier frequency currents along a long telephone circuit.

Shunt conductance is also expressed as the reciprocal of resistance or $G = 1/R$ mhos. However, it is usual to indicate this loss in ohms per loop mile as for resistance.

Line Characteristic Impedance

The previously described distributed constants of a long telephone line, that is, its series resistance, series inductance, shunt capacitance, and leakage, determine its *characteristic impedance,* Z_0. This is one of the most important operating characteristics of a telephone transmission line. The value of the characteristic impedance, Z_0, is determined by the values of the aforementioned distributed constants of the line. In general, this means that the characteristic impedance is largely determined by the type of cable facility. It is entirely independent of the length of the cable and is not affected by the impedance of the connecting terminal equipment to the line. The importance of the characteristic impedance to the telephone plant is that it imposes specific requirements upon the equipment connected to the telephone line, both at its transmitting and receiving ends. That there may be maximum transfer of power to and from the telephone line, it is necessary that the connecting equipment have the same characteristic impedance as that of the cable conductors.

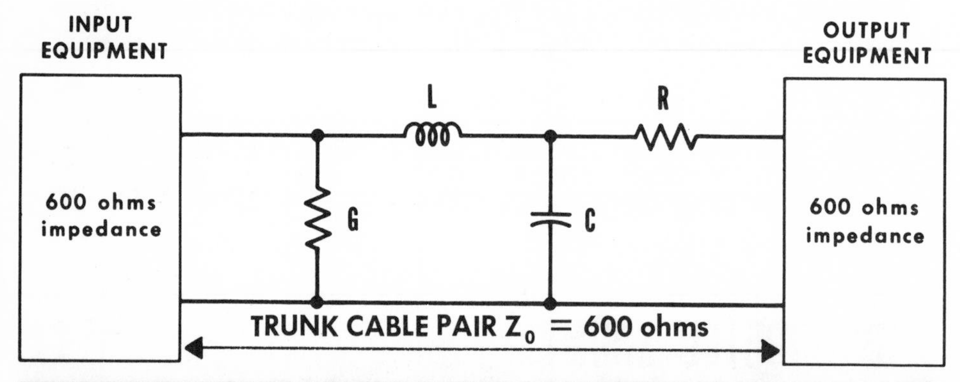

TELEPHONE LINE WITH 600-OHM CHARACTERISTIC IMPEDANCE, AS REPRESENTED BY THE DISTRIBUTED LINE CONSTANTS L, C, R, AND G

The distributed line constants of a cable pair, particularly its inductance (L) and capacitance (C), cause its characteristic impedance (Z_0) to vary slightly with frequency. For instance, the nominal Z_0 of cable pairs used for interoffice and toll connecting trunks is assumed to be 600 ohms at 1,000 Hz. Cables which interconnect subscriber lines to the central office are designed for 900-ohms characteristic impedjnce (Z_0). The Z_0 will be slightly greater at higher frequencies and less at lower frequencies. These impedance differences normally are not sufficient to cause difficulties to the operation of carrier systems over the cable pairs. It is essential, for the previously explained reasons, that the impedance of carrier transmitting and receiving terminals be equivalent to that of their assigned cable pairs.

Attenuation in Transmission Line

Speech waves traveling along a long telephone line gradually lose their strength because of *attenuation* in the line. This is a simplified manner of stating that there is a gradual diminishing of the magnitudes of their instantaneous current and voltage amplitudes. One reason for this is the voltage drop (I × R) along the length of the line caused by its distributed series resistance. Another is that the leakage, G, and shunt capacitance, C, between the conductors prevent some of the current from traveling the entire distance. The power, which is the product of the voltage and current values (E × I), will gradually decrease with distance from the transmitting end of the line

To understand this action better it is good engineering practice to divide the long telephone circuit into *line sections*, each of a specified unit length. For example, a 50-mile (31.2 km) line can be divided into five sections of 10 miles (6.25 km) each. A line section would contain finite values of the four main electrical properties of the line, namely, R, L, C, and G. The amplitudes of the voltage and current waves will decrease at the same rate over the entire line length only if the aforementioned distributed constants have uniform values in each line section.

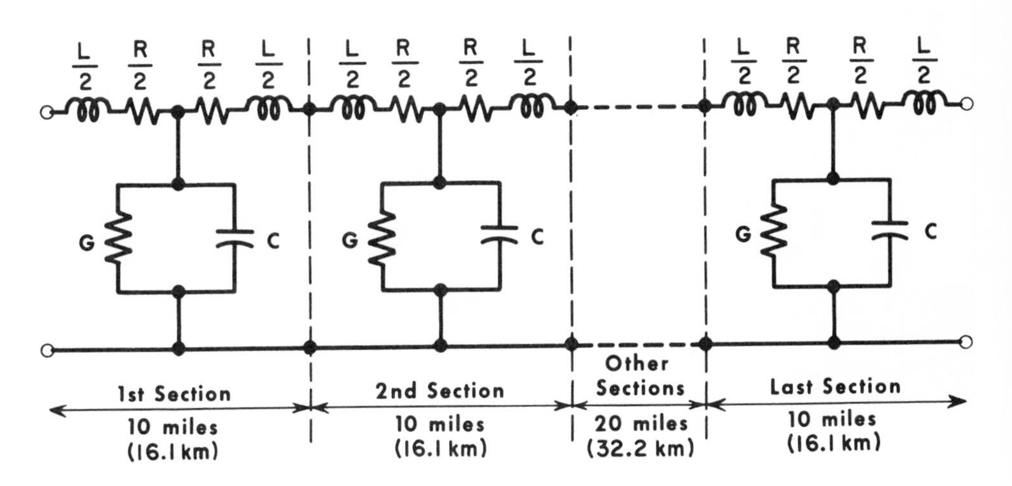

LINE-SECTION REPRESENTATION OF 50-MILE-LONG TELEPHONE CIRCUIT

Where the line sections have different distributed constants, the characteristic impedance, Z_0, will be changed. This will result in a different rate of decrease in input and output current and voltage values in each line section. The resultant output, therefore, will not decrease uniformly along the line and certain voice frequencies will be reduced, or attenuated, more than other frequencies. This may cause distortion in the speech signals received at the end of the line unless certain corrective measures are taken at the terminal equipment.

Phase Velocity

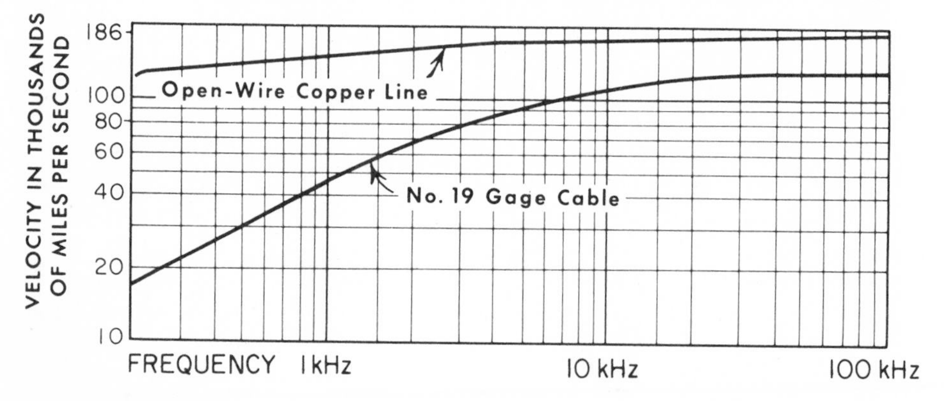

Electricity travels at the speed of light. This velocity of propagation is approximately 186,000 miles per second or 300,000,000 meters per second. On open-wire telephone lines and coaxial cables, the velocity is almost equal to the speed of light for frequencies above about 4,000 Hz. In telephone cables such as the standard No. 19 gage toll type, however, the propagation velocity varies greatly with frequency. For example, the velocity is about 16,000 miles per second (25,600 km/sec) at 100 Hz, increasing to 45,000 miles per second (72,000 km/sec) at 1,000 Hz, and to about 110,000 miles per second (176,000 km/sec) at 10,000 Hz. At a frequency of 50,000 Hz, it becomes almost constant at around 125,000 miles per second (200,000 km/sec). Speech signals, as we learned before, comprise a complex wave of frequencies in the 200-3,400-Hz range. Waves of these different frequencies do not move along the telephone line with the same velocity. Waves of the highest frequency travel with the greatest velocity as discussed above. This is because of the skin effect phenomenon referred to in our discussion on series resistance. It reduces the inductance of the wires and, therefore, the inductive reactance, X_L, at the higher frequencies.

The velocity of the speech wave as it moves along the line is called the *phase velocity.* The speed of propagation along the line is determined by the line characteristics. In the obsolete open-wire line, the velocity varied from about 140,000 miles per second (224,000 km/sec) at 250 Hz to almost the speed of light at 3000 Hz. However, if the line is a cable pair, the velocity will range from about 25,000 miles per second (40,000 km/sec) to 75,000 miles per second (120,000 km/sec) for the same band of frequencies. It is apparent that the various waves in the 200-3,400-Hz range which comprise speech travel with different velocities in the cable. It should be noted, however, that carrier waves which usually range above 4,000 Hz (4,000 to 1,550,000 Hz) travel at about the same constant velocity on open-wire lines or in cable pairs.

Echo or Reflection Effects

We have learned that the characteristic impedance, Z_0, of a long telephone line should remain the same for all sections. If not, there will be *reflections* at the points where the impedance changes. This could result from such cases as connecting a cable pair to a circut with larger gage wire, changing the cable gage at a line section to a smaller size, or terminating the line in equipment having different impedance values. When the voice or carrier wave reaches the impedance mismatch point, as for example at the direct connection of an underground cable pair to a different gage circuit, reflection will take place. In other words, not all the power will be transferred from the first to the second cable pair. A portion will be reflected back along the telephone line to the transmitting terminal causing an *echo* effect.

The power which is sent back will be in the form of voltage and current waves. Therefore, the telephone line will contain one pair of voltage and current waves moving to the receiving end and another pair traveling back to the transmitting terminal. The closer the first cable pair matches the second pair's impedance, the less will be the reflected or lost power. This points out the importance of reducing the reflection or echo effects on long telephone lines—that is, of increasing the return loss—because:

1. The power delivered to the terminal equipment is reduced.
2. The resulting reflected wave or echo effect distorts the wave shape, thereby degrading speech quality.

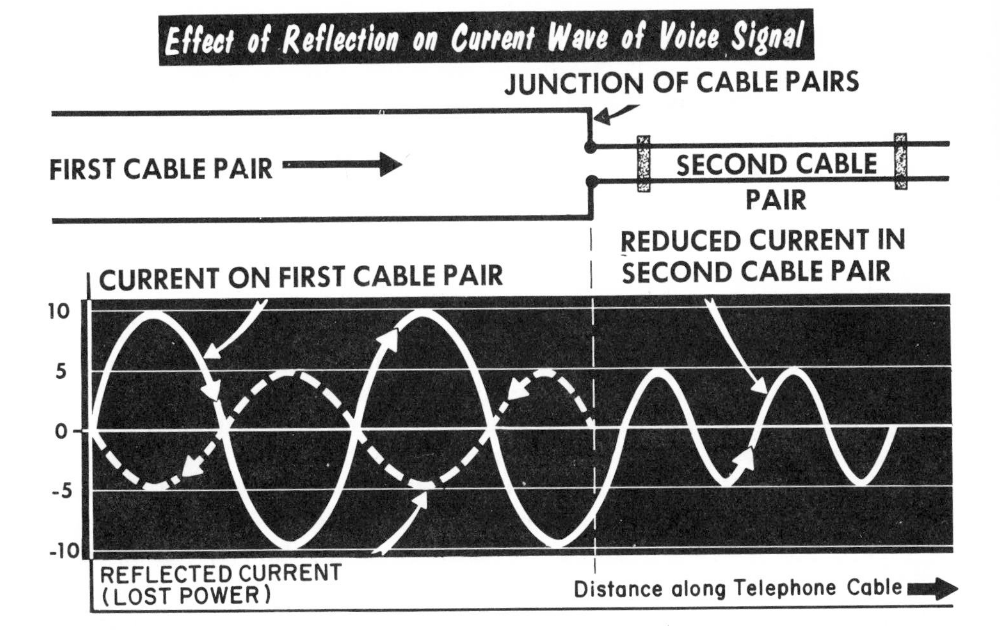

Transmission Measurements

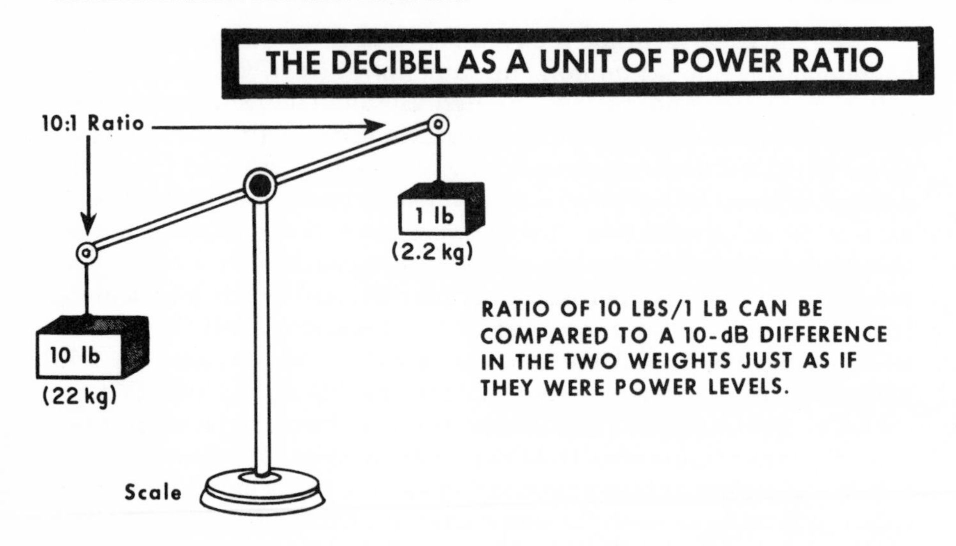

In an electrical distribution system, the power losses can be readily calculated from measurements made by instruments such as voltmeters and ammeters. The frequency in power systems is fixed at 60 cycles per second (60 Hz). In a telephone system, the power levels are extremely low and the frequency is not constant. Therefore, it is not practical to utilize such instruments for making measurements. A person talking into a telephone, for example, produces a voice-frequency signal of a very small fraction of a watt—about 0.0001 watt. The attenuation in, say a 10-mile (6.25 km) line, probably will reduce the signal to around 0.000001 watt at the receiving end.

Now, let us insert a repeater in the line to extend the talking range. The signal may be amplified about 100 times. The power level would be increased back to 0.0001 watt. Similarly, if hybrid coils, filters, or other networks were inserted in the telephone line, additional power losses would take place. To measure and numerically express such power loss as well as the effects of crosstalk, distortion, repeater gain, etc., telephone engineers in the United States many years ago developed a system of transmission measurements. The system is based on measuring the input and output signal powers of a telephone circuit and expressing their ratio in terms of a unit called a *decibel.* The original unit was called the Bel after Alexander Graham Bell, the inventor of the telephone. By definition, 1 Bel is equal to a power ratio of 10 to 1. This proved to be too large a unit for practical use. It was decided, therefore, to use a unit equal to 0.1 Bel or the decibel. The decibel, abbreviated dB is a unit of power *ratio,* not of power. It can be used to indicate either loss or gain of power change. It is customary to use a minus sign to signify a power loss (as -4 dB) and a plus sign to show a gain (as $+$ 6 dB or 6 dB).

The Decibel

The decibel is used widely in telephone and radio communications and in the electronics field to express logarithmically the ratio between two values of voltage, current, or sound levels as well as power. It is used often to express the amount of power loss in a telephone line, hybrid network, or the gain of an amplifier. The decibel is also defined as the unit of attenuation caused by 1 mile of standard No. 19 gage cable at a frequency of 886 Hz. As a logarithmic unit, the decibel closely indicates the response of the human ear to sound waves. The human ear responds approximately proportional to the logarithm of the energy of the sound wave, and not to the energy itself. The decibel as a unit also permits losses and gains in the components of a telephone line or system to be added or subtracted by simple arithmetic to obtain the resultant numerical expression of the overall gain or loss.

To understand completely the decibel system of measurement requires a knowledge of the common system of logarithms used in mathematics. This is based on the power or exponent of the number 10. It is beyond the scope of this book to treat and explain this mathematical system and the method of using logarithms to express numbers. However, the general formula for determing a ratio, in decibels, of two power levels is shown in the illustration.

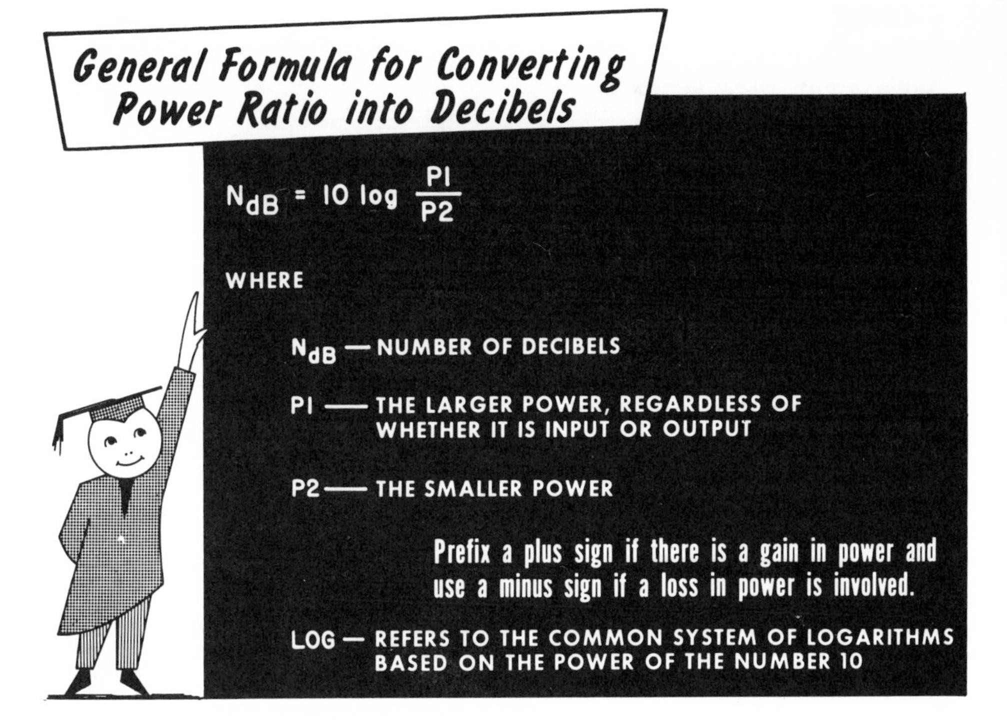

The Decibel (cont.)

Fortunately, it is not necessary to make extensive calculations and to refer to a table of logarithms each time we desire to compute the power ratio in terms of decibels or vice-versa. We can refer, instead, to a chart which gives power ratios as well as current and voltage ratios directly from decibels and, of course, vice-versa. The following is one form of such chart (sometimes termed a *dB table)* commonly used in engineering work. It is adequate for all normal usage of decibels and power and voltage levels that may be required in carrier telephony or associated work.

DECIBELS AND POWER, VOLTAGE, OR CURRENT RATIOS

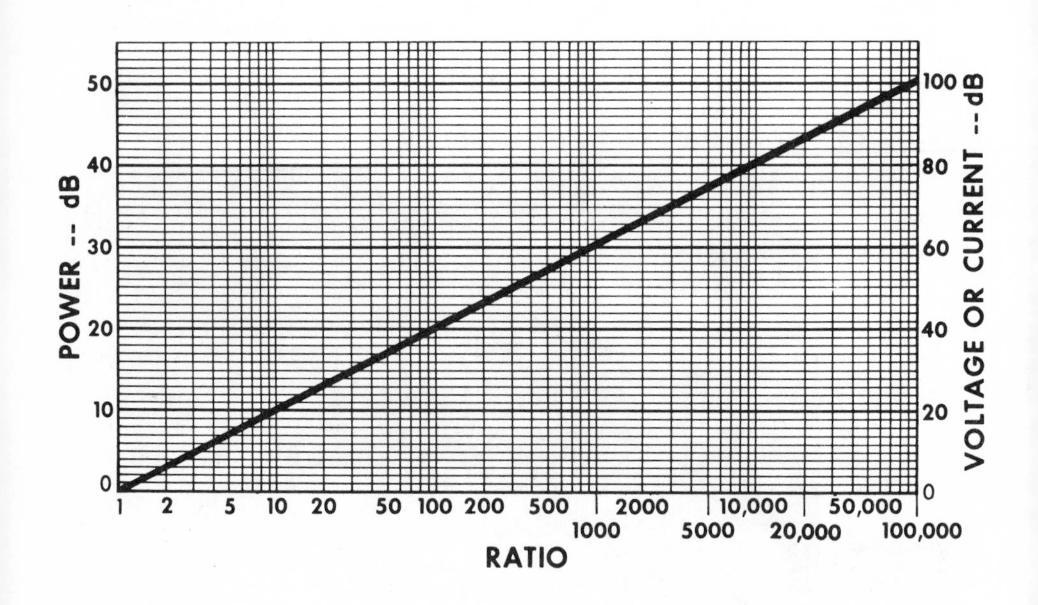

Use of the Decibel

The following examples on the use of the *decibel* refer to the telephone transmission applications. The same principles may be applied to radio transmission systems, sound levels in audio-frequency amplifiers, and many other applications in the electronics field.

1. *Power gain:* A repeater has an input power of 0.002 watt and produces an output of 20 watts. What is the power gain in decibels (dB)? Referring to the dB table, we see that a power ratio of 10,000 is equivalent to a gain of 40 dB.

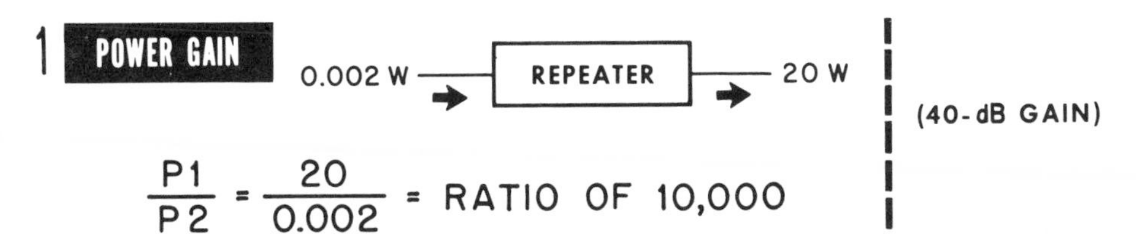

2. *Power loss:* The power of a 1,000-Hz tone from a test set at a toll center trunk is found to be 0.001 watt at the connection to a toll trunk. The power input at the distant center termination of the trunk measures 0.000005 watt. What is the power loss in the line in dB? Referring to the dB table, a power ratio of 200 equals a loss of −23 dB. The minus sign is used to indicate that this is a power loss.

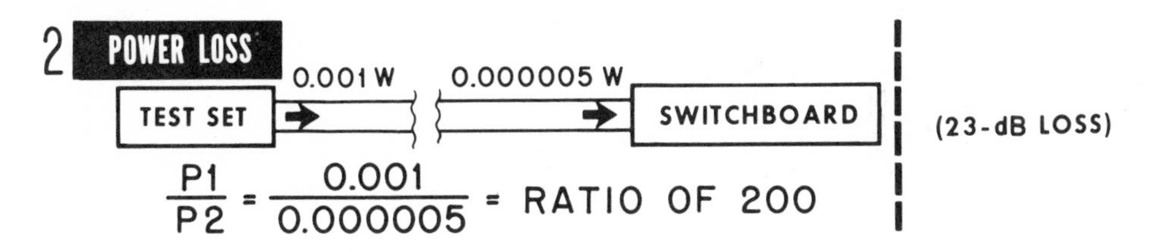

3. *Voltage loss:* The dc voltage of a line at the MDF (main distributing frame) measures 48 volts. At a distant subscriber's station, the voltage is found to be 12 volts. What is the loss in dB? From the voltage ratio part of the dB table or chart, we see that a voltage ratio of 4 is equivalent to −12 dB.

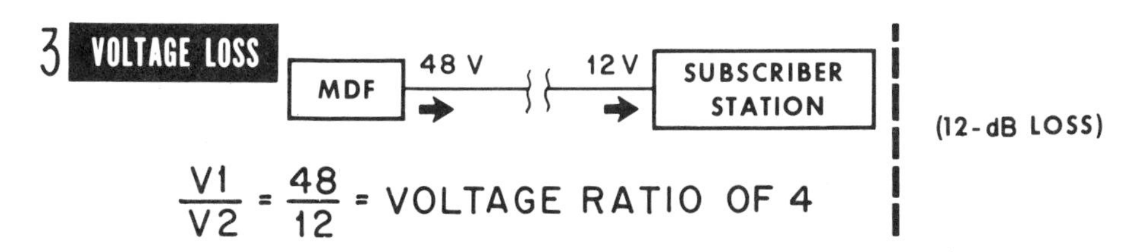

Use of the Decibel (cont.)

4. *Power ratio:* A repeater having a gain of 20 dB is connected to a 10-mile (6.25 km) long telephone line. This line is found to have a loss of approximately 4 dB per mile. A 1,000-Hz test tone at a power of 0.001 watt is sent into the repeater. What is the power in watts at the end of this 10-mile (6.25 km) telephone line?

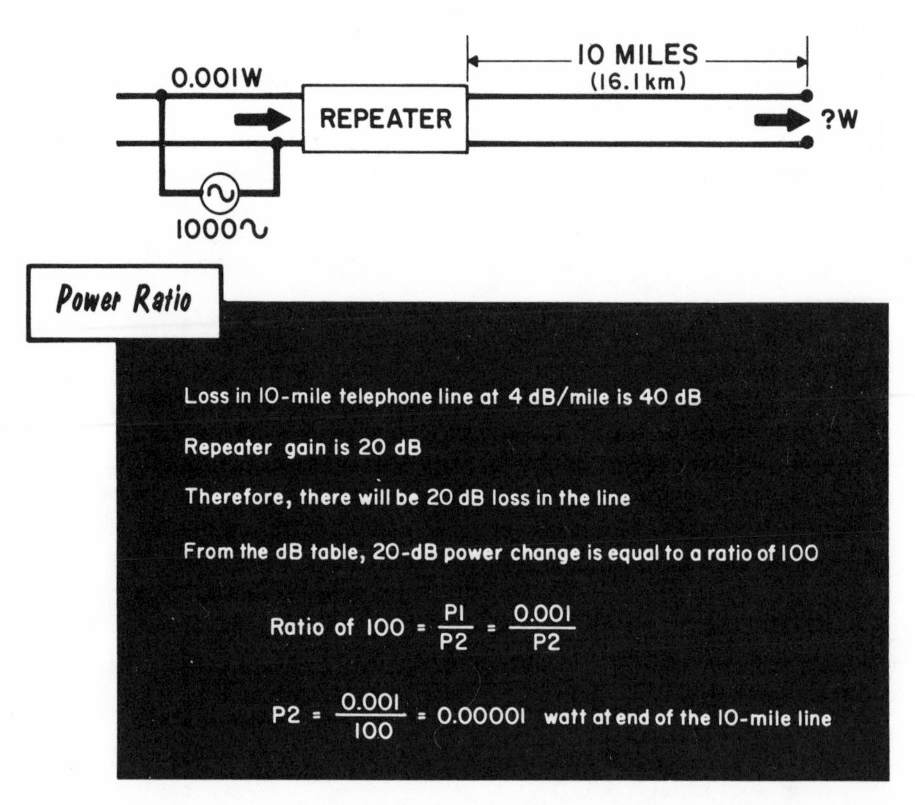

It is often desirable to use the decibel as an absolute unit in telephone transmission measurements. For this purpose, it has become the practice to assign a *zero level* fixed reference value to which ratios may be referred. This zero level is 0.001 watt or 1 milliwatt and is termed 0 dBm (where m refers to 1 milliwatt). As an example, the output power of a 10-watt amplifier in a repeater can be expressed in decibels as follows:

$$\text{Power Ratio} = \frac{10}{0.001} \text{ watt, where } 0 \text{ dBm} = 0.001 \text{ watt}$$

This gives a power ratio of 10,000, which from the dB table or chart is equal to 40 dB. We can say, therefore, that the 10-watt amplifier provides a gain of 40 dB.

Questions and Problems

1. What are the electrical properties of a telephone line and their symbols?
2. What is "skin effect" and what distributed constant is mainly affected by this phenomenon?
3. What is the formula for the inductive reactance of a cable pair?
4. What is shunt conductance?
5. What determines the characteristic impedance of a telephone cable?
6. Why is the characteristic impedance of a cable pair so important?
7. Do speech currents travel at the same velocity as carrier currents over a cable pair?
8. What causes reflections and what are their effects?
9. What transmission measurement unit is used in communications? Define it.
10. If a telephone circuit has an attenuation of 30 decibels, what is the power level of a repeater inserted into the circuit to provide a 0-dBm level at the distant end of the line?

Transmission Mediums

The different transmission mediums generally utilized for the transmission of telephone and for other forms of electrical communications are: (a) multiconductor cables, either suspended on poles or underground; (b) underground and undersea coaxial cable systems; (c) point-to-point HF, VHF, and microwave radio systems; (d) satellite radio circuits; and (e) open-wire pole lines.

The open-wire pole line was the original telephone transmission medium, just as it had been for telegraph messages. It has now been almost entirely replaced by cable and radio facilities. Well constructed open-wire toll lines had low losses at speech and carrier frequencies relative to cable circuits. Let us briefly review the electrical characteristics of former open-wire lines, as shown below. This should aid us in understanding the development of telephone transmission standards and their applications to cables and other transmission means.

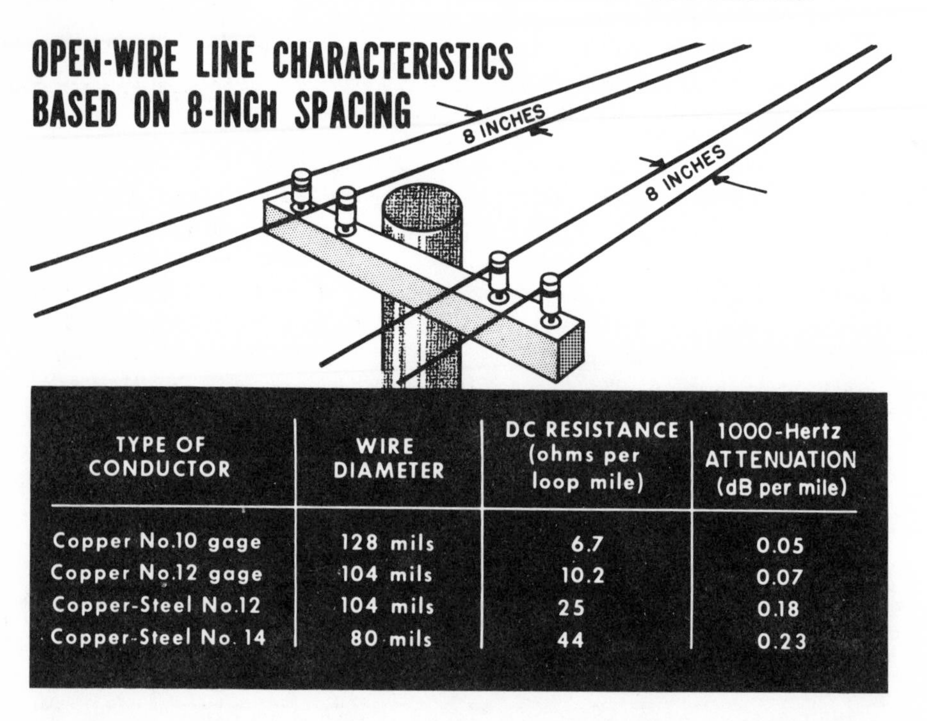

TYPE OF CONDUCTOR	WIRE DIAMETER	DC RESISTANCE (ohms per loop mile)	1000-Hertz ATTENUATION (dB per mile)
Copper No.10 gage	128 mils	6.7	0.05
Copper No.12 gage	104 mils	10.2	0.07
Copper-Steel No.12	104 mils	25	0.18
Copper-Steel No. 14	80 mils	44	0.23

Weather conditions greatly affected the attenuation of open-wire lines. The resistance of the wire increases with temperature. Wet and humid conditions increase the attenuation of the line. Copper-clad steel wires were normally utilized for improved mechanical strength. They had about 40% of the conductivity of all-copper wire of the same size. However, at frequencies above about 1000 Hz, copper-steel wire had about the same attenuation as all-copper wire because of the skin effect.

Line Interference and Crosstalk

On a pole line, telephone wires are adjacent to each other and are near the ground surface. As a result, they are susceptible to *noise* and *crosstalk* interference. The noise may come from parallel power lines which normally carry high voltages at 60 Hz. Harmonics of this 60-Hz frequency can fall within the 200-3,400-Hz voice range and cause noise on the telephone circuits. Electromagnetic or inductive coupling between adjacent telephone circuits on the same pole line also produces interference. This interference is called crosstalk. It causes the conversation on one circuit to be faintly audible on another line.

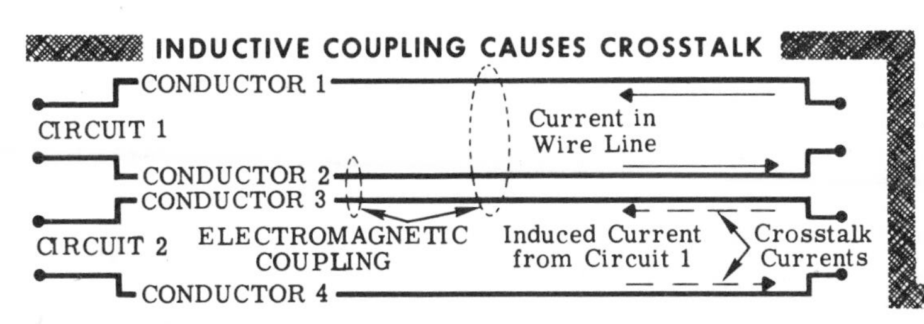

Since conductor 3 is closer to conductor 2 than is conductor 4, there will be different voltages induced in conductors 3 and 4. This potential difference causes the flow of crosstalk currents.

It is essential that noise and crosstalk be held to a minimum on telephone lines so that speech can be transmitted with the least loss of intelligence. Noise from power lines can be kept at a low value by adequate separation between telephone and power wires, and by maintaining a balanced condition between each conductor of the telephone line and ground. Crosstalk between adjacent pairs of wires can be held to a minimum by the proper spacing of the conductors. The separation between the two conductors of a pair must be kept small in relation to the distance between adjacent pairs. As a rule, the conductors of an open-wire toll line were spaced 8 inches apart, while adjacent pairs were separated 16 inches.

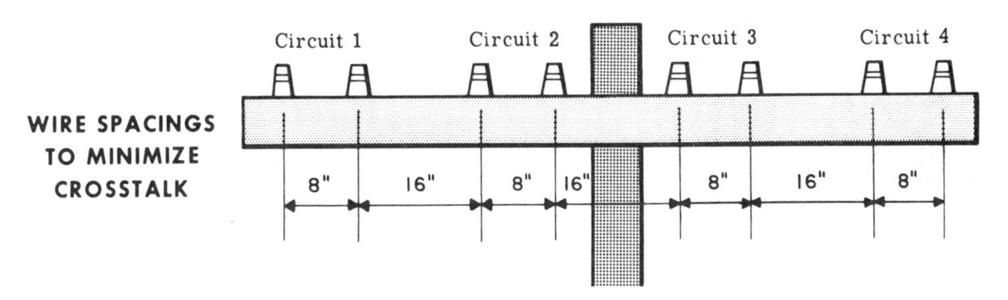

Close spacing caused both wires of a pair to receive about the same amount of induced voltage from an adjacent pair. The resultant crosstalk, therefore, was very small. Large separation between adjacent pairs reduced the magnitude of the induced voltage, minimizing the crosstalk.

Transposition of Open-Wire Lines

A pole line with only a single pair of similar wires far removed from power lines or other sources of electrical interference is an ideal condition. There are, however, usually several wire pairs on a telephone pole line. The power lines are nearby or even on the same pole. The two conductors of a particular telephone pair cannot be equally distant from all other wires on the pole line. As a result, different voltages are induced in the two conductors of this pair from speech transmissions on an adjacent pair or from the power lines. *Transposition* is a method of minimizing this interference on open-wire lines. It involves the interchange of the positións of the two conductors of each pair at specific intervals along the telephone line. To be effective, each pair of conductors must be transposed at electrically short intervals for noise reduction. To reduce crosstalk between two adjacent pairs, one pair must have at least one additional transposition more than the other pair.

Transpositions are usually made by either the point or rolling methods. In the point method, the two conductors of a pair are interchanged within a space of a foot or less. A point type transposition bracket utilizing four insulator pins can be provided for this purpose. This is widely used on carrier circuits. The line wires always remain parallel to each other except at the transposition bracket.

POINT TYPE TRANSPOSITION

WIRE 1
WIRE 2
WIRE 2
WIRE 1
Transposition Bracket on Crossarm

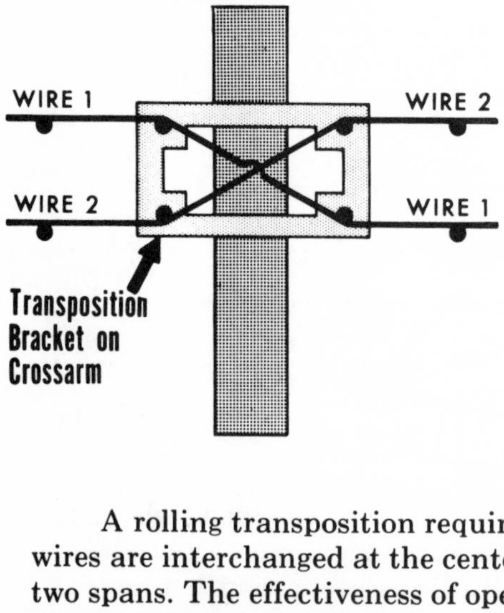

ROLLING POINT TRANSPOSITION

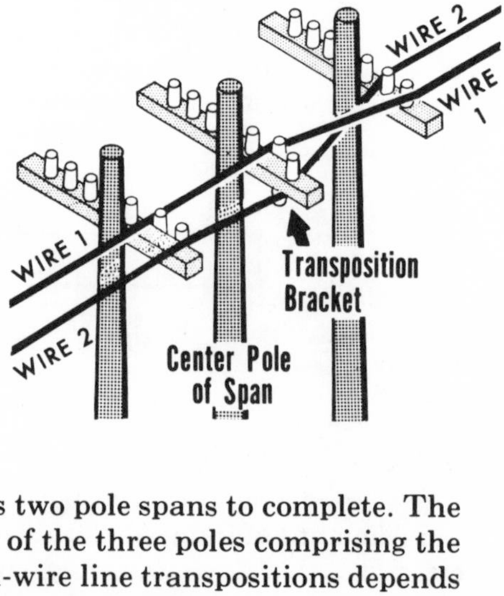

WIRE 2
WIRE 1
WIRE 1
WIRE 2
Transposition Bracket
Center Pole of Span

A rolling transposition requires two pole spans to complete. The wires are interchanged at the center of the three poles comprising the two spans. The effectiveness of open-wire line transpositions depends upon a constant line impedance. Bad splices, faulty insulation, etc., must be avoided to minimize crosstalk and noise interference.

Cable Characteristics

Cable and microwave circuits have now replaced almost all open-wire lines, even in most rural areas. Coaxial cables employing high-capacity carrier systems, such as those described later, now provide many of the long distance circuits. Multiconductor cables are used for certain toll and long distance circuits, for interoffice trunks, and for connecting customer lines to central offices. Bell System statistics reveal that in 1954, open-wire facilities provided about 14% of the toll and long distance circuits; cables provided 68%; and point-to-point radio only 18%. Now, more than twenty years later, the percentages are less than 0.3% for open-wire, 36% for coaxial cables, and 64% for microwave radio including satellite circuits.

Multiconductor cables were originally lead covered and contained paper-insulated wires. Plastic-insulated wire cables are used almost exclusively at present. To minimize crosstalk and other interference effects, cable conductors are continuously transposed by intertwining the two wires of each pair. In subscriber cables, conductors are usually No. 22, 24, 26, or 28 gage respectively. Trunk and toll cables normally employ No. 19 gage wires, although No. 16 gage may be used. In these cables, two pairs with identical electrical characteristics are frequently intertwined to form a four-wire group or *quad*. One pair transmits in one direction; the other pair transmits in the opposite direction. This arrangement, also known as a *four-wire circuit* is often utilized for toll circuits with various cable carrier systems.

Attenuation losses in cables are much higher than for open-wire lines. The greater capacitance between conductors of a cable pair is the main cause of this higher attenuation. These losses increase considerably with frequency. Carrier systems were thus first developed for open-wire lines. Typical cable characteristics are shown below.

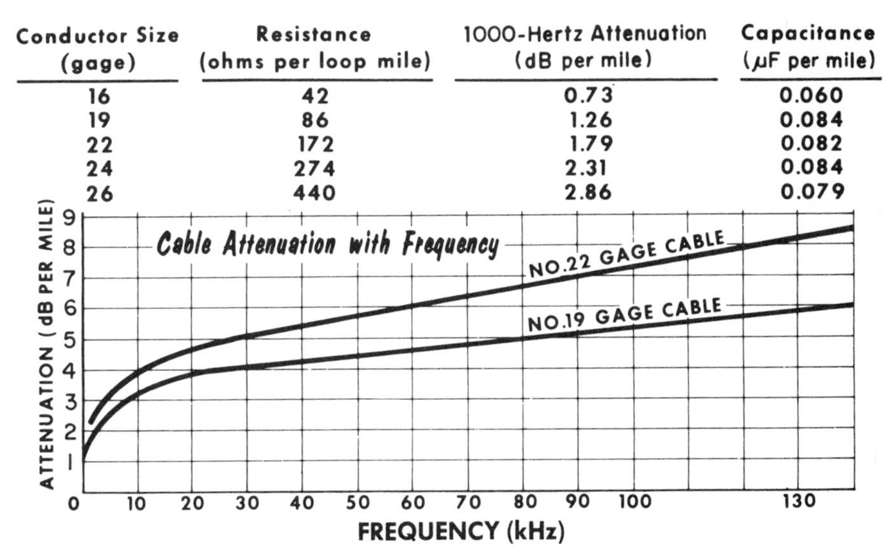

Conductor Size (gage)	Resistance (ohms per loop mile)	1000-Hertz Attenuation (dB per mile)	Capacitance (μF per mile)
16	42	0.73	0.060
19	86	1.26	0.084
22	172	1.79	0.082
24	274	2.31	0.084
26	440	2.86	0.079

Loading of Cable Circuits

Cable circuits, as we have learned, possess much higher losses than open-wire lines. The size of the cable conductors could be increased to reduce the distributed resistance and thereby lower the attenuation. This would, however, materially decrease the number of circuits per cable. Increasing the separation between the conductors of a cable pair reduces the distributed capacitance. This also decreases the attenuation of the cable. But, the number of circuits in the cable is substantially reduced.

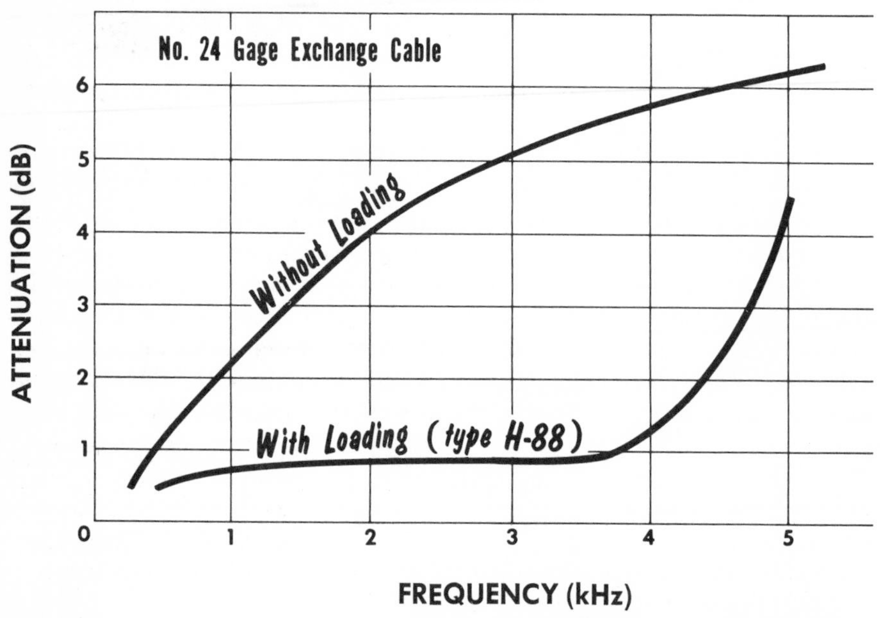

The best solution to decreasing the attenuation in a cable has been found to be in an economical way to increase its distributed inductance. This method is called loading. The effect of loading a cable circuit is to change its characteristics. By proper loading, the characteristic impedance and the attenuation per unit length of a cable can be held constant over the desired voice-frequency range. This will also cause the maximum transfer of power to and from the cable circuit over this range of frequencies. The cable-loading method was developed for voice-frequency circuits. It is not utilized in present cable carrier systems, as explained later.

Loading Coils

Loading Coil Construction

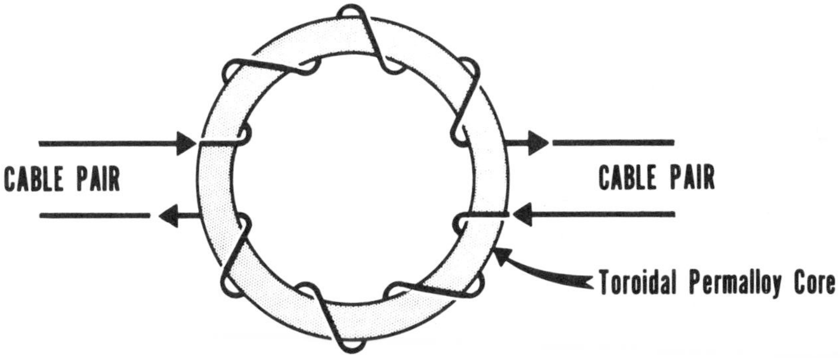

CABLE PAIR

CABLE PAIR

Toroidal Permalloy Core

1. The two windings are placed in opposite directions around the core.
2. Current flow is in opposite directions through these windings.
3. The magnetic field of both windings is in the same direction around the core. The changing magnetic field with the current reversals adds inductance to the circuit.

The distributed inductance of a cable pair can be increased uniformly by wrapping it with a steel or permalloy tape. This is a rather expensive way. The most practical and economical solution is to insert loading coils at equal intervals along the cable pair. The spacing of the loading coils must be calculated to give the same effect as if the wires were uniformly wrapped with the aforementioned magnetic material. A loading coil may be described as a doughnut-shaped or toroidal core wound with copper wire. This core is usually made of powdered permalloy to provide a high magnetic permeability. There are air pockets in the core material to reduce the possibility of saturation, thereby preventing distortion of the signals.

The loading of a cable circuit must be designed for the particular use of the transmission facility. For example, one type of loading is used for entrance cables that connect to open-wire lines and another type for toll cable circuits. For transmitting carrier frequencies, the loading coils must have smaller inductance values and be installed at closer intervals than for voice frequency currents. Incidentally, it should be specifically noted that *open-wire lines are never loaded.* The attenuation characteristics of open-wire lines are not constant because of the adverse effects of weather conditions. This makes it impractical to utilize loading coils to increase the line's inductance as a means of improving the line's efficiency.

Principles of Loading

To better understand the principles of loading as applied to telephone cable circuits, let us consider a mechanical analogy. In the illustration, a piece of light string, say about 10 feet (3.04 m) long, is shown fastened to a hook in a wall. It can be compared to a cable pair. By a snapping movement of the hand, waves of motion are generated in the string. They tend to die out rapidly before traveling to the end of the string. If a heavier string were used, the waves would be propagated a little longer distance along the string before dying out. The heavier string may be considered analogous to a larger-gage wire which has a lower resistance.

Let us tie a small lead weight, like that used on a plumb line, to the center of the string. The waves created in the string by the snapping movement of the hand will again decrease rapidly. In fact, the waves will cease when they reach the weight. This illustrates that a large amount of loading cannot be added to a circuit at one point. Now, let us take about 20 small wooden beads or similar light weights and space them equally along the entire distance of the string and tie them. When the string is now snapped, the resultant wave motion will travel the entire distance of the string to the wall hook with only a slight decrease in amplitude. This analogy shows that proper loading reduces the attenuation and distortion in a cable circuit.

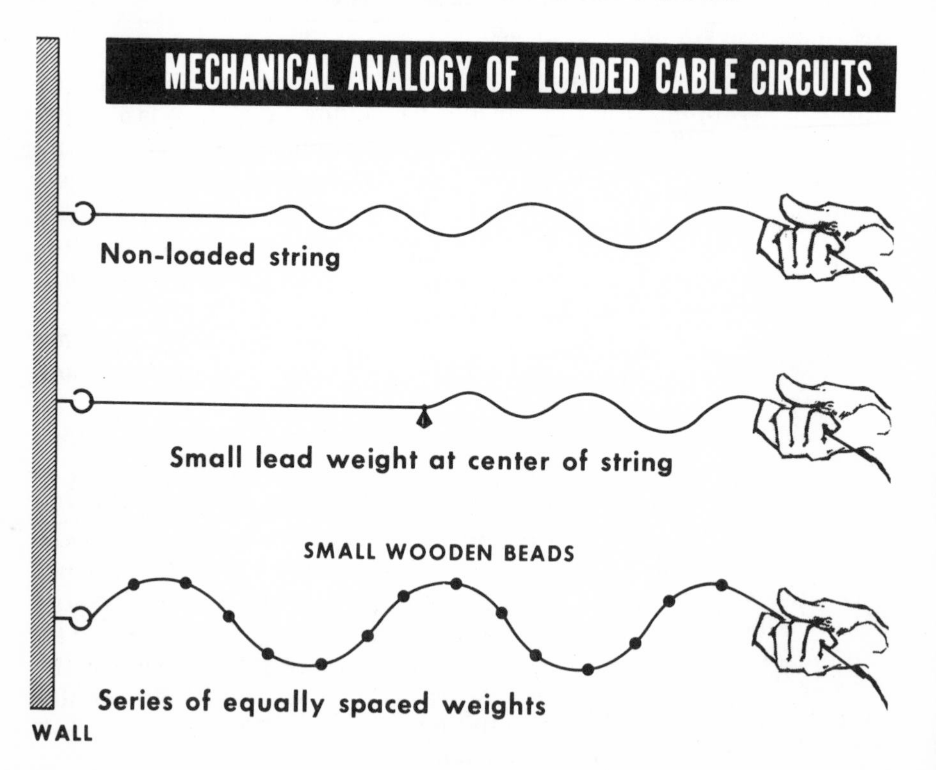

Loading Limitations

LOADED CABLE CHARACTERISTICS AT 1000 Hz						
TYPE OF LOADING	VELOCITY (miles per sec)	LOADING COIL SPACING	CUTOFF FREQUENCY (Hz)	ATTENUATION (dB per mile)	DISTRIBUTED CONSTANTS PER LOOP MILE	
					Inductance (mh)	Resistance (ohms)
No. 19 gage toll cable						
NONE	46,900	------	-----	1.06	1	84
H-31	23,300	6000 ft	6,700	0.56	28	87
H-44	20,000	6000 ft	5,700	0.49	39	88
H-88	14,300	6000 ft	4,000	0.36	78	91
H-172	10,300	6000 ft	2,900	0.28	151	96
B-88	10,200	3000 ft	5,700	0.28	156	98
No. 24 gage exchange cable						
NONE	25,000	------	-----	2.15		
M-88	14,900	9000 ft	3,100	1.31		
H-88	12,700	6000 ft	3,700	1.13		
B-88	9,270	3000 ft	5,300	0.86		
No. 22 gage exchange cable						
NONE	29,400	------	-----	1.80		
M-88	14,500	9000 ft	2,900	0.92		
H-88	12,100	6000 ft	3,500	0.79		
H-135	9,800	6000 ft	2,800	0.63		
B-88	8,750	3000 ft	5,000	0.60		
B-135	7,050	3000 ft	4,000	0.48		

The combination of the distributed capacitance and the lumped inductance of the loading coils causes a loaded cable circuit to act like a bandpass filter. This effect greatly increases the attenuation of the cable when the cutoff frequency is exceeded. This cutoff frequency is determined by the capacitance of the cable and the inductance of the loading coils. By installing proper loading coils and spacing them at certain prescribed intervals, the cutoff frequency can be increased beyond the upper end of the frequency band to be transmitted.

There is a practical limit to the number of loading coils that can be inserted into a cable circuit. Each loading coil increases the series resistance and the inductance of the cable conductors. This added resistance could soon overcome the beneficial effects of loading. The tabulated data show some typical characteristics of loaded cable circuits. Note that the conductor resistance increases and the cutoff frequency decreases as the inductance value of the loading coil is raised.

Loading, furthermore, reduces the travel speed of cable voice and carrier currents because of the added inductance. Although loading has many advantages for voice-frequency circuits, it will adversely affect carrier systems that operate above the cutoff frequency of the loading scheme. Cable pairs for carrier circuits above 15 kHz are normally not loaded.

Phantom Circuits

In voice-frequency networks, the conductors of two adjacent circuits may be utilized to form a third circuit with the aid of repeating coils. This derived third circuit is called a *phantom circuit.* The combination of the two physical or *side* circuits used to form the phantom circuit, and the phantom circuit, is called a *phantom group.* A simplified schematic diagram of the phantom group is shown.

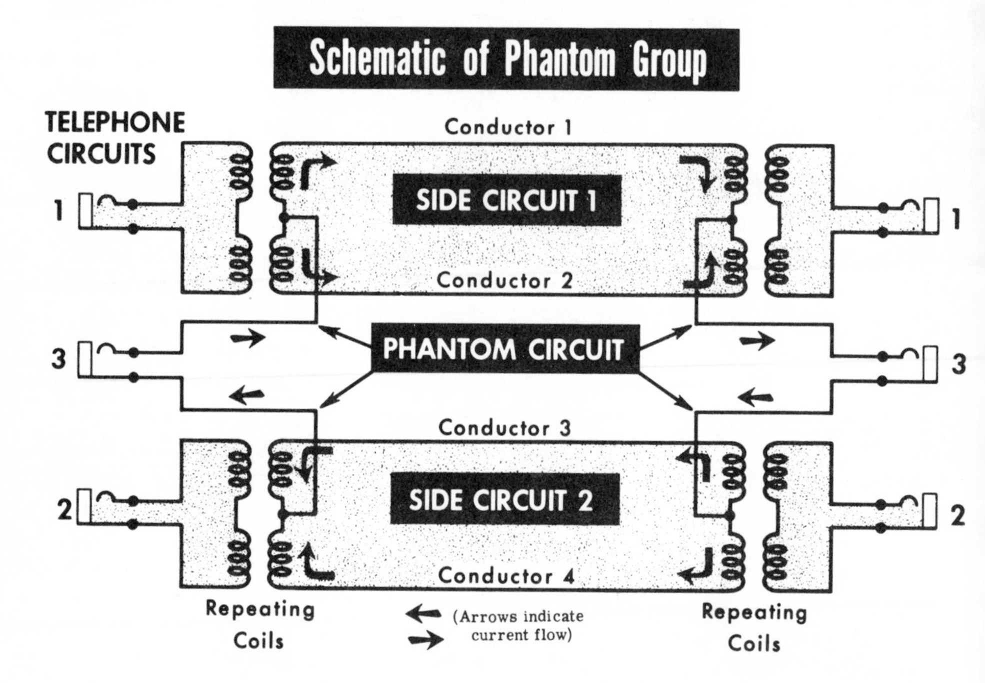

The two wires of side circuit 1 serve as one conductor of the phantom circuit. Similarly, the two wires of side circuit 2 act as the other conductor of the phantom circuit. Three telephone circuits are obtained in this manner from only four conductors. It is essential that the repeating coils used be properly balanced, that is, the two halves of their secondary windings should have equal turns. This is necessary so that the voice currents of the phantom circuit can divide equally between conductors 1 and 2 as well as between conductors 3 and 4. In this way, there will be no difference of potential across the repeating coils in side circuit 1 caused by the current in the phantom circuit. Therefore, no interference will be caused to the two side circuits by the phantom circuit, and vice-versa. It is important that the resistance of each conductor of side circuits 1 and 2 be the same. Otherwise, the resistance unbalance will make unequal currents flow in the conductors. This will result in crosstalk interference between the phantom and side circuits. Phantom circuits, as a result, are not suitable for carrier transmissions.

Telephone Signaling

In addition to transmitting conversations or messages, a telephone circuit must be able to handle a variety of signaling functions. For instance, certain supervisory signals are used to indicate that one central office desires to establish a connection to another. Signals are also used to pass information on the progress of a call or the status of a circuit. Different techniques are employed for signaling over local subscriber's circuits than for long distance trunks. One of the earliest signaling methods employed the hand-operated magneto generator at the telephone station. By cranking it, a 16-20-Hz ac voltage was generated to ring the bells. It is still used for special purposes.

Throughout the years, 20-Hz ac (100-105 volts) has become the standard ringing voltage in telephone systems. In certain PBX operations it is still used but only for manual ringdown signaling. The operator presses a key in the cord circuit of the PBX switchboard to send the ringing current as a signal to the distant switchboard. Sometimes 20-Hz, 135-Hz, or 1,000-Hz signaling was used on intertoll trunks. Ringdown signaling using 1,000 Hz will go through voice-frequency repeaters. It is necessary, however, either to bypass the 20- and 135-Hz signals around the repeaters or to convert them to 1,000 Hz. The latter or voice-frequency ringing method, which amplifies the ringing current, is the preferred arrangement. Current signaling methods will be discussed later .

An Early Telephone with Magneto Generator

Trunk Signaling

The electrical energy for signaling in the modern telephone plant is usually supplied in the form of: direct current from the central office battery; alternating current of a single frequency; or multiple-frequency tones. Signaling may be required in both directions of transmission, as on toll trunks, sometimes simultaneously and often sequentially. There are two conditions of particular importance in the general use of signals for interoffice trunks. They are broadly described as *on-hook* and *off-hook* supervision. These terms originated from the action of the receiver on the switchhook of the early *candlestick* type of telephone. With the receiver on the switchhook, the line was disconnected and the on-hook condition prevailed. When the receiver was removed, as to make a call, the condition changed to off-hook.

In general, the on-hook signal is indicative of the idle condition at the distant or far end of the trunk. This type of signal is also transmitted on a change from the busy to the idle state. The off-hook signal signifies a busy condition at the far end of the trunk. These and other signals required for supervisory and control purposes may be transmitted over the trunk in several ways. Direct-current signaling is generally used for interoffice and short-haul toll trunks without repeaters. Alternating-current single-frequency tones are usually used for signaling over carrier systems and for long distance trunks with voice-frequency repeaters.

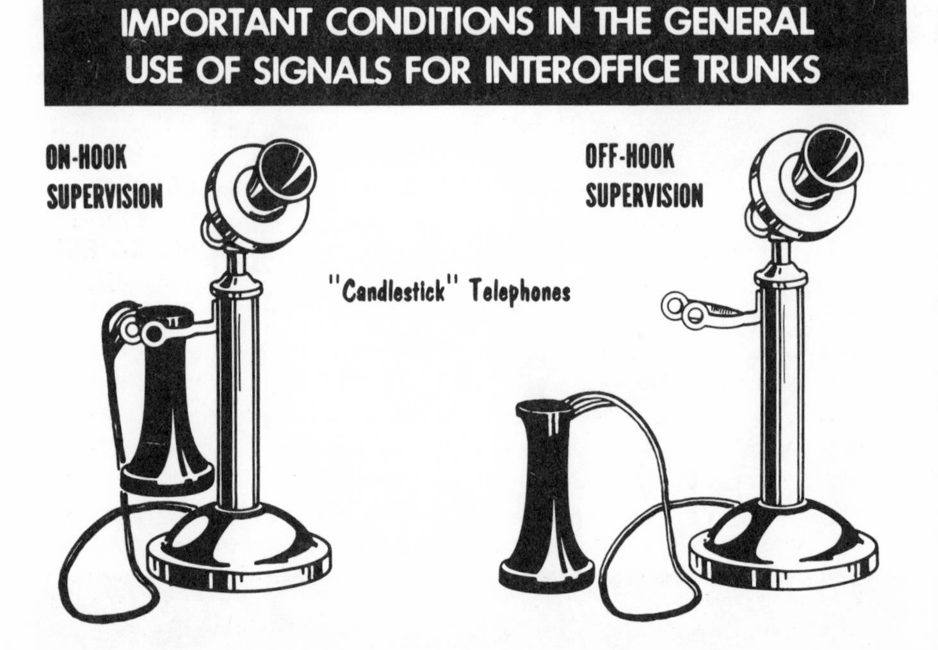

E & M Signaling System

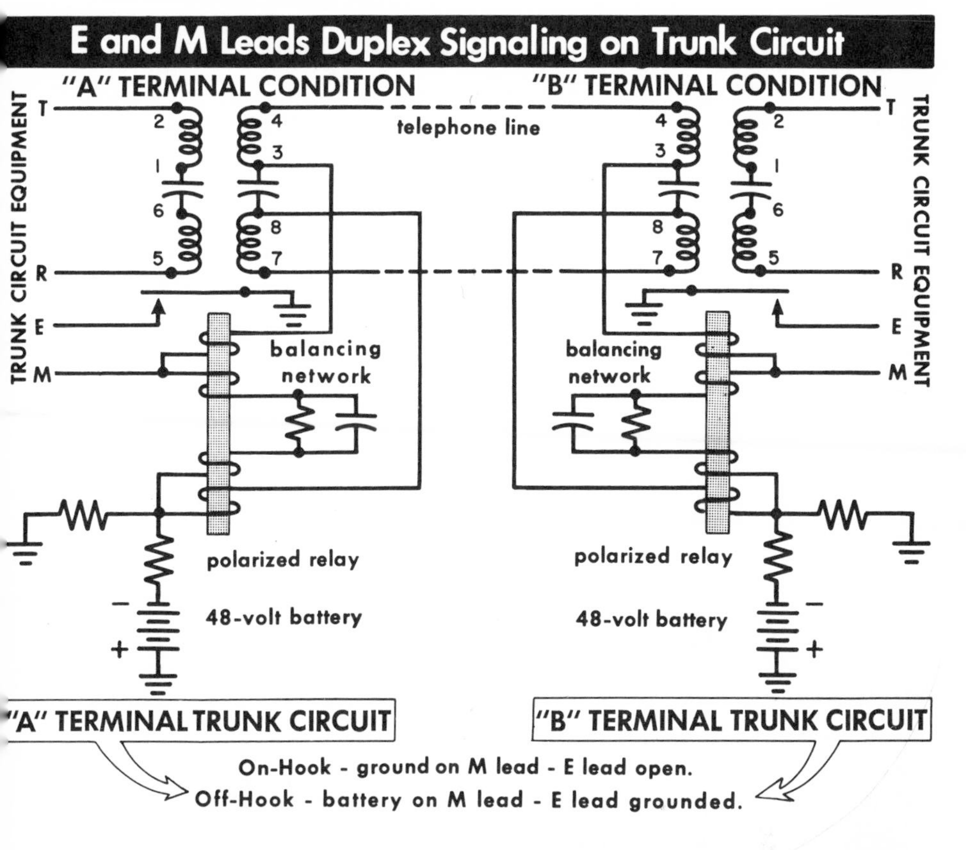

Communication between a trunk circuit and its associated signaling system is normally accomplished over two leads. They are, by convention, designated the *E* and *M* leads or conductors. The M lead transmits battery or ground signals to the signaling system. The E lead receives open or ground signals from the trunk. In other words, the near-end condition of the trunk is reflected by the M lead and the far-end condition by the E lead. The system of E and M lead signaling received its name from certain historical designations used on old circuit drawings.

The schematic diagram shows a method of transmitting on-hook and off-hook conditions by the E and M lead signaling system. It is also possible to arrange connections between ringdown signaling and E and M lead signaling systems. The equipment used for E and M lead signaling is usually separate from the trunk equipment. Various circuit arrangements such as simplex, duplex, composite, and reverse battery have been devised to permit the E and M system of signaling between central offices on a dc basis. It is also applicable to ac signaling over carrier systems.

Power Levels and Voice Frequency Repeaters

We have learned that the attenuation of telephone transmission lines limits the distance over which one can converse. So that the received voice signals may have sufficient power to actuate the receiver in the telephone handset, amplification often is necessary. Voice-frequency amplifiers are normally employed for this purpose. Amplifiers in a circuit arrangement of balancing networks and hybrid coils comprise a *voice-frequency repeater.* A simplified amplifier circuit in a repeater is shown in the illustration. Note that the *point* of the triangle in the symbol of the amplifier always points in the direction of the output.

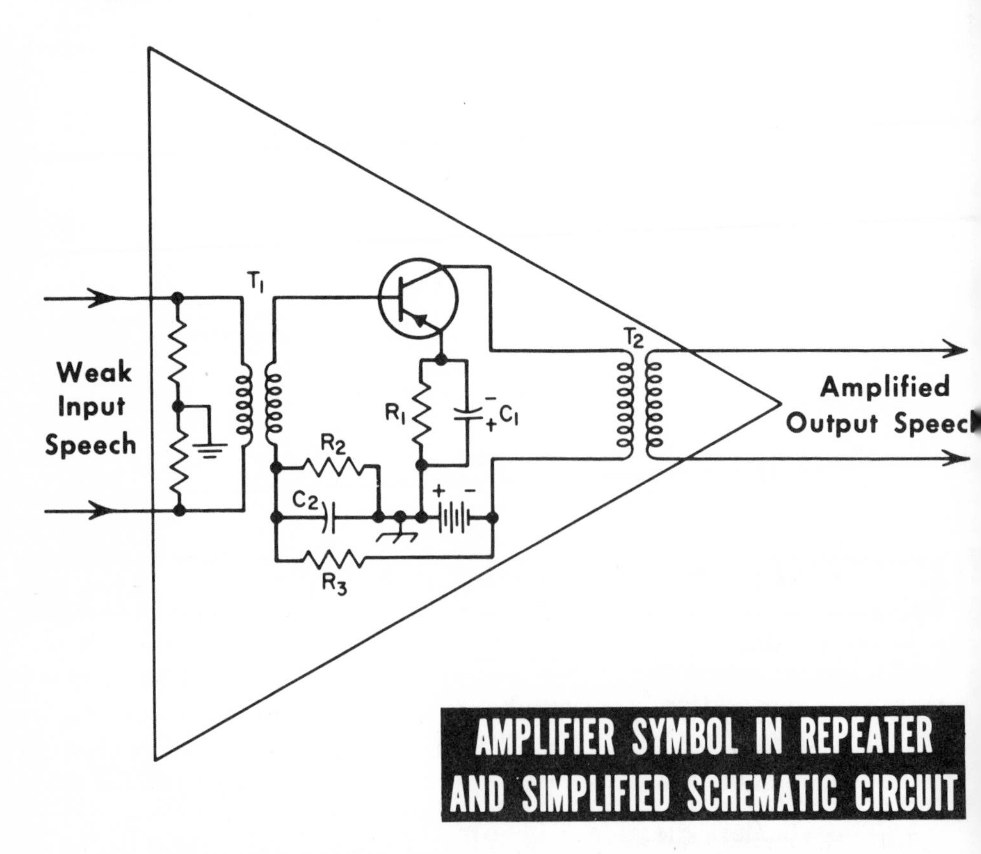

AMPLIFIER SYMBOL IN REPEATER AND SIMPLIFIED SCHEMATIC CIRCUIT

The average telephone receiver requires a power level of about −30 dBm (decibels referred to 1 milliwatt in 600 ohms). This means that a signal with approximately one-millionth (0.000001) of a watt power is the lowest level that will produce an audible sound from the receiver. Amplifiers designed for use in repeaters can be adjusted to give power gains from about 6 dB to 30 dB. The important part played by voice-frequency repeaters in a long telephone circuit may be better understood from a study of a *power-level diagram.*

Power Level Diagram

The following is an example of a power-level diagram for a two-wire toll circuit. It shows one-way transmissions from toll center A to toll center B.

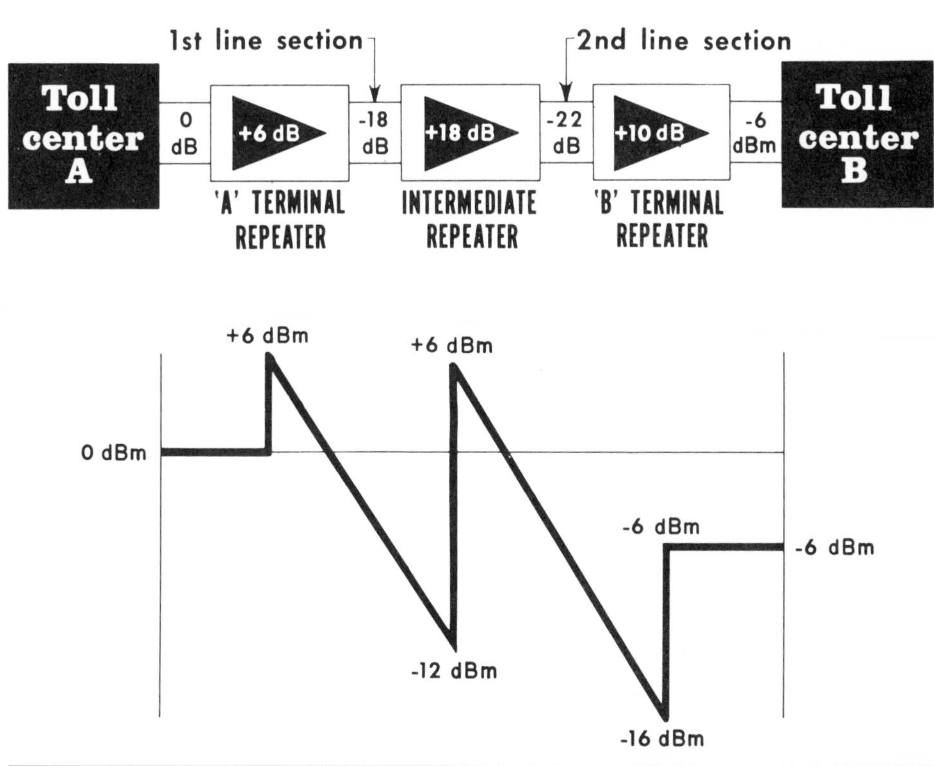

POWER-LEVEL DIAGRAM FOR ONE-WAY TRANSMISSIONS

It indicates the power level in dBm (0 dBm = 0.001 watt in 600 ohms) in sections of a two-wire telephone circuit containing voice frequency repeaters. Starting at toll center A, the first amplifier raises the voice power level to +6 dBm at the output of the terminal repeater. There is a loss of 18 dB in the first line section. This loss reduces the power level of the signal to −12 dBm at the input of the intermediate repeater. The net gain of this repeater is set at 18 dB. This raises the signal power to +6 dBm at the output of the intermediate repeater. Note that the second line section has a loss of 22 dB. Therefore, the resultant input into the terminal repeater of toll center B is −16 dBm. There is a net gain of 10 dB in this terminal repeater which provides a −6 dBm signal level at the B toll center. This power level is adequate for connection to subscribers normally served directly by toll centers or through another central office.

Questions and Problems

1. What transmission mediums are used for telephone communications? Which one was used originally?
2. What causes crosstalk on open-wire telephone circuits?
3. What is the purpose of transposition on pole lines? What two main types are used?
4. Why are attenuation losses in cables much higher than for open-wire lines?
5. How may the attenuation of voice-frequency currents in a cable pair be decreased economically?
6. How does a loading coil add inductance to a cable pair? Are open-wire lines loaded? Why?
7. What are the adverse effects of loading cable conductors? Is loading used with carrier systems?
8. What is the advantage of a phantom circuit? Is it used with carrier systems?
9. What are the important signaling conditions for trunk circuits? What designation of the signaling system is usually employed?
10. What does a power-level diagram portray?

Frequency Translation in Analog Carrier Systems

We have learned that a telephone line transmits direct current and, theoretically, can also transmit an infinite number of alternating current frequencies. In practice, however, frequencies up to about 150,000 Hz may be effectively transmitted over open-wire lines. Wire pairs in multipair cables, likewise, can convey a similar range of frequencies, but with greater attenuation. Since speech signals are normally in the 200-3,400-Hz band range, only a very small part of the wire pair's available frequency range is being utilized. For this individual voice band assignment or space division method, it would be necessary to provide a separate pair of conductors for each telephone channel, including interoffice trunks and intertoll network circuits. This inefficient and expensive way of furnishing more telephone channels led to the development of carrier transmission systems.

Generally, carrier systems may be classified as analog or digital types. Analog carrier systems, which were developed first, are designed to transmit the continuous waveform version of the input voice signal. Analog systems usually use the Frequency Division Multiplex (FDM) technique with amplitude *(AM)* or frequency modulation *(FM)*. Note, the FDM method has many voice-frequency channels (200-3,400-Hz range) and they are spaced out on the same pair of conductors by means of the frequency translation technique. This permits a large number of conversations to be transmitted at the same time without mutual interference. The drawing below illustrates three conversations simultaneously carried over the same wire pair.

The Time Division Multiplex (TDM) arrangement utilizes a sequence of time units or slots controlled by electronic gates, to convey samples of the voice signals. At the TDM carrier receiving terminal, the original voice signals are reconstructed for transmission in the initial analog manner. The TDM technique is used primarily in Pulse Code Modulation (PCM) carrier systems, which will be described later. Digital carrier systems, such as the PCM type, transmit trains of pulses. These streams of pulses and spaces (no pulses) represent analog signals (input speech) which have been encoded into a series of binary numbers. They are transmitted over the cable pair in designated time units and are decoded to recover the original analog voice signals at the receiving PCM carrier terminals.

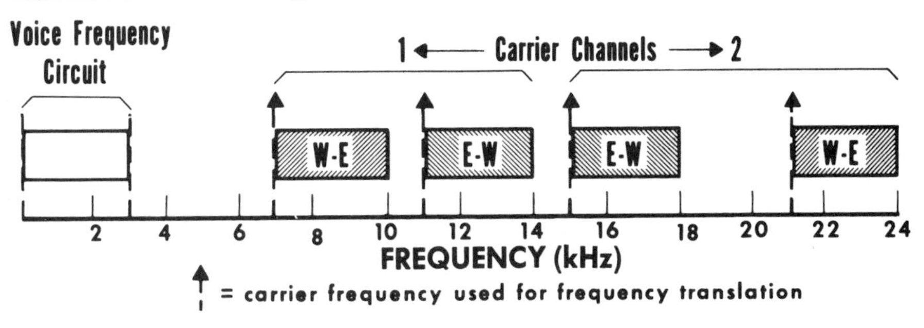

Carrier Channel Signals

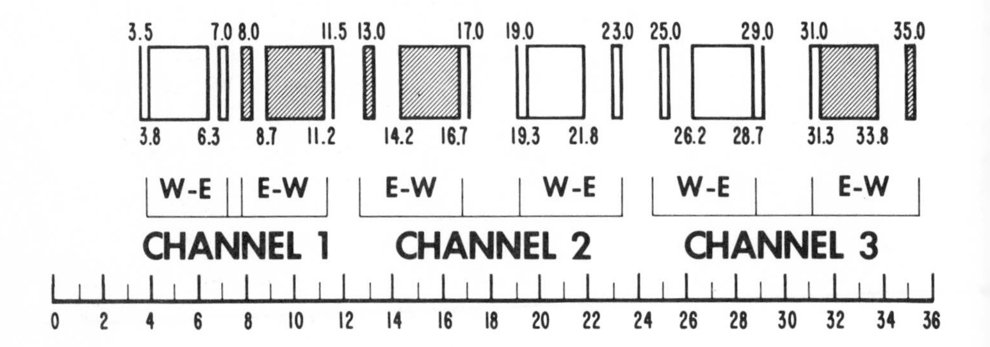

In FDM carrier systems, the carrier signal is an alternating current (ac) of a predetermined and constant frequency. It is generated within the carrier terminal equipment by solid-state oscillators instead of the vacuum tubes used in early carrier systems. A separate carrier frequency is required for translating the voice-frequency signals sent to each carrier channel. Different carrier frequencies are needed for each direction of transmission on the same wire pair. The frequency of the carrier oscillator signal must be above the voice band (200-3,400 Hz) for frequency translation operations, and to prevent interference with any voice-frequency channel.

The drawing shows an example of the carrier channel frequencies provided in early 3-channel carrier systems used on open-wire lines. Carrier signals were generated at 6.8, 20.8, and 28.3 kHz respectively for transmission in one direction. In the other direction, the generated carrier frequencies were 10.3, 16.8, and 32.8 kHz for channels 1 to 3, respectively. The designations, *West to East* (W-E) and *East to West* (E-W) are commonly used to identify terminal points and their associated equipment units in carrier systems. Since most carrier frequencies are in thousands of Hz, it is also customary to indicate the frequency range in kilohertz (kHz). One kilohertz (kHz) is equal to 1,000 hertz (Hz).

Carrier Frequency Allocations

The carrier oscillator frequencies, as discussed, form the basis for the frequency translation operations. This is accomplished by the modulation process which we will study later. Each carrier channel must be assigned to a different frequency range for simultaneous transmission over the same wire line. The proper selection of the frequency range for the various carrier channels is essential to the efficient operation of a carrier system. This is also quite important for coordination purposes to prevent interference between different carrier systems on the same or adjacent open-wire lines or cable pairs. This process is termed *frequency allocations.* It includes provision for a small frequency separation between adjacent carrier channels as a sort of guard band. This frequency separation requirement results from the filters used in carrier systems. The filters are designed to pass only the frequencies assigned to a particular carrier system and to attenuate all others. The frequency bands, therefore, of the different carrier systems on the same wire pair cannot be the same or overlap.

The chart shows the frequency band allocations of some typical carrier systems used in subscriber and toll cables. The Lenkurt #47A and W.E. type N1 carrier systems are of the double sideband type; while the Lenkurt #46B and W.E. type N3 utilize the single sideband suppressed carrier mode of transmission. The single sideband method, discussed later, permits twice the number of channels to be transmitted within the same frequency band. Note that different frequency bands are used for opposite directions of transmission.

LENKURT #82A 1 TO 6 CHANNEL STATION CARRIER
(SINGLE WIRE PAIR)

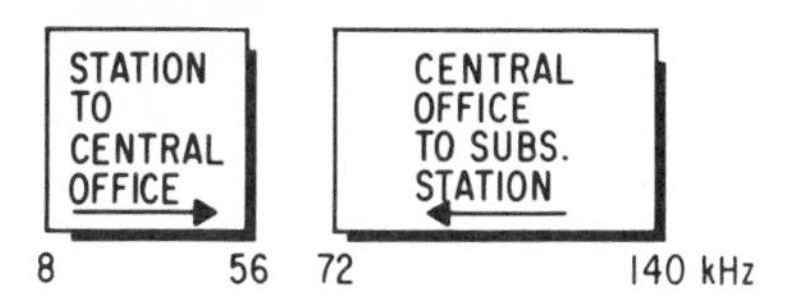

TYPICAL CARRIER SYSTEM FREQUENCY ALLOCATIONS

LENKURT #47A OR WESTERN ELECTRIC TYPE N1
(12 CHANNELS ON TWO CABLE PAIRS)

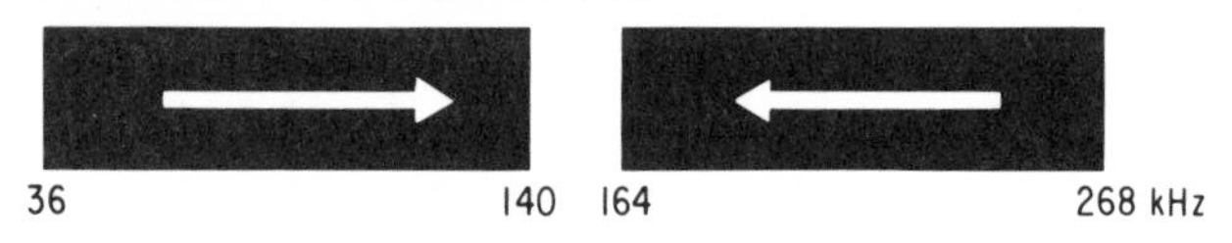

LENKURT #46B OR WESTERN ELECTRIC TYPE N3
(24 CHANNELS ON TWO CABLE PAIRS)

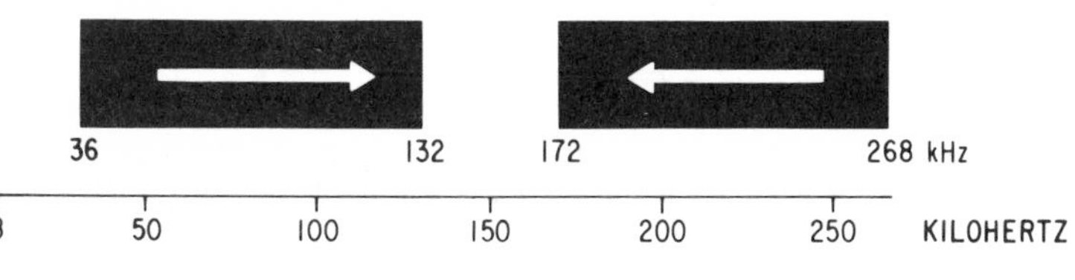

Frequency Coordination

In addition to the frequency allocation problem for carrier systems on the same cable or open-wire pair, there is also the case of two carrier systems operating on separate cable or open-wire pairs. Their frequencies are physically separated but there will be interference because of inductive coupling. This is due to the two or more cable or open-wire pairs paralleling each other in the same cable or on the same pole line, respectively. As a result of this condition, some of the carrier signal power in one pair will be induced in, or transferred to, the adjacent pair. This could cause considerable crosstalk between the two carrier systems. It is desirable, therefore, to also coordinate frequency allocations on an interpair basis to minimize the crosstalk. The greatest interference usually occurs at points where one system transmits a much higher signal level than the other system. This often exists at carrier terminals and repeaters.

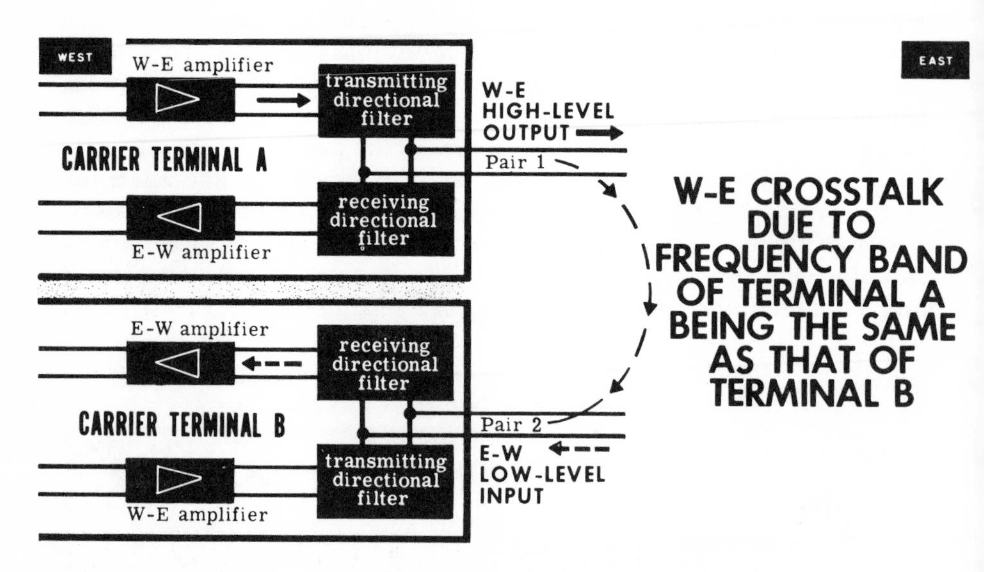

In the diagram, the W-E output of carrier terminal A is at a high level. The signals entering carrier terminal B from the E-W direction are at a much lower level due to the line attenuation. If the W-E frequency band of terminal A were the same as the E-W band of carrier system B, the crosstalk from system A would pass through the E-W amplifier of terminal B and cause interference. By using a different W-E frequency band for carrier system A than for that of system B, the directional filters of carrier terminal B will be able to attenuate the crosstalk. In essence, interpair coordination requires that the W-E band of one carrier system differ from the E-W band of the other system for operation on the same pole line or in the same cable. Carrier equipment manufacturers have been coordinating carrier frequency allocations for many years to avoid these difficulties.

Frequency Inversion and Staggering of Carrier Channels

Crosstalk between parallel open-wire and conductor pairs, employing the same carrier systems, may also be effectively reduced by *frequency inversion and staggering* techniques. By the proper choice of carrier oscillator frequencies in the channel modulators and demodulators, as will be described later, it is possible to obtain either the upper or lower sideband. These sideband products of modulation have been discussed briefly under "Telephone Carrier Elements." The upper sideband is also termed the *upright* sideband, and the lower sideband the *inverted* sideband.

In the upper or upright sideband, the voice signals occupy the carrier channel space in the normal sequence of 200-3,400 Hz. When the lower or inverted sideband is used, the voice signals are inverted to appear in the frequency order of 3,400-200 Hz. Therefore, by frequency inversion of carrier channels on adjacent cable pairs, any resultant crosstalk will also be inverted in frequency. Such inverted crosstalk is unintelligible and appears as noise to persons using carrier systems on adjacent lines. Frequency inversion, consequently, may reduce crosstalk interference by 50% between similar carrier systems on adjacent conductor pairs.

The frequency band of a carrier system may be *staggered,* or shifted slightly, with respect to similar carrier systems on adjacent open-wire pairs. This is also accomplished by the appropriate selection of carrier oscillator frequencies in the respective modulator and demodulator circuits. Any crosstalk into nearby conductor pairs, likewise, will be unintelligible and resemble noise.

The illustration shows frequency inversion and staggering techniques previously used in the GTE Lenkurt 45A carrier system.

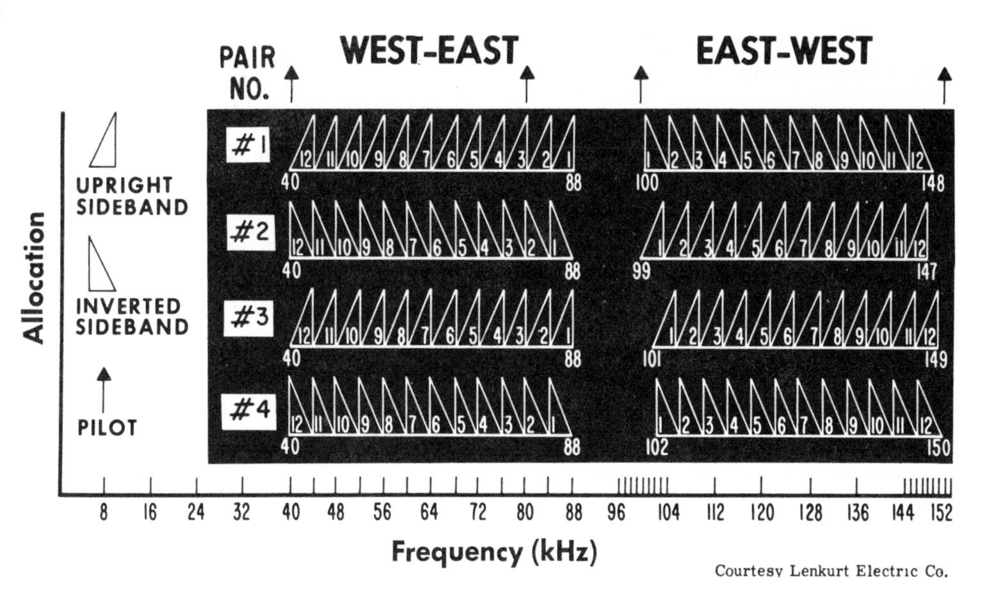

Courtesy Lenkurt Electric Co.

Carrier Channel Basic Elements

Each carrier channel comprises certain basic elements or components. These components are in addition to the carrier frequency generator or oscillator previously discussed. They are essential parts of both the transmitting and receiving sections of each carrier channel. These basic items are the: carrier frequency generator or oscillator; modulator and demodulator (also called *modem*); bandpass, low-pass, or other filters; and transmitting and receiving amplifiers. The block diagram illustrates their functions in a typical single-channel carrier channel used for two-way conversations.

BASIC ELEMENTS OF A CARRIER CHANNEL

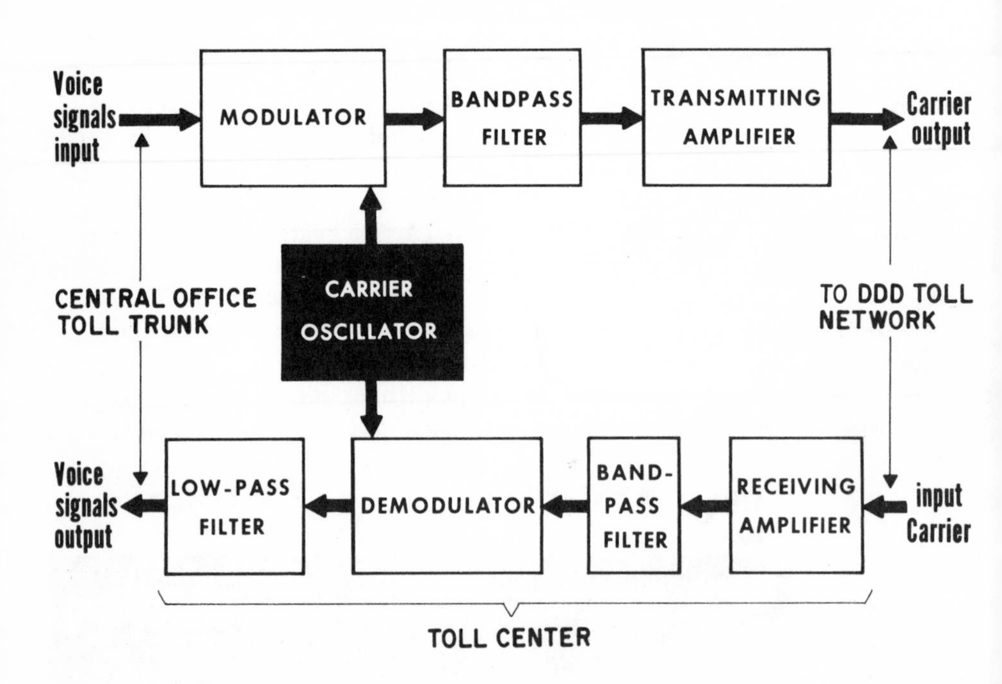

Several such carrier channels can make up a *carrier terminal.* The purpose of each basic carrier element and the important part that it plays in providing carrier telephone communications will be discussed in more detail in the suceeding pages. Note that other components such as hybrid coils, balancing networks, pads, limiters, and other items are employed in a complete carrier terminal. They are needed to provide for two-wire and four-wire connections and other operating features. Our discussions, however, will be limited at this time to the above basic circuit elements.

Modulator Functions

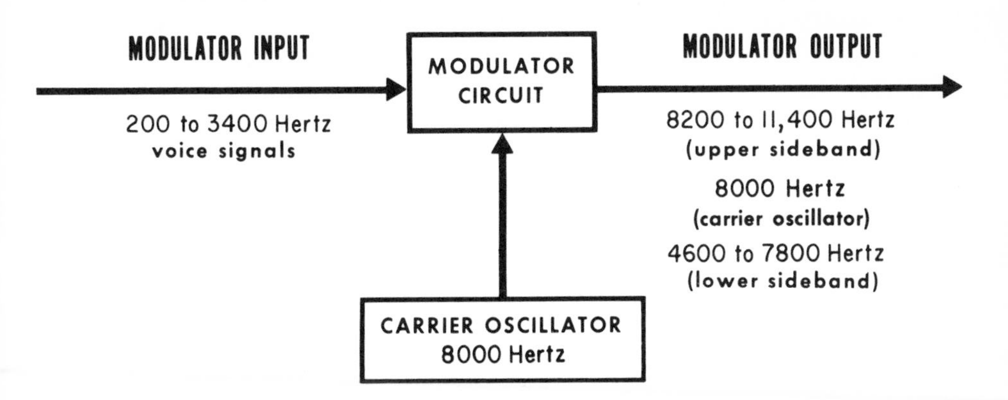

To understand better the frequency translation and other functions of the modulator, let us follow the transmission of voice signals through it. Speech or voice frequency signals, which are in the 200-3,400-Hz range, enter the modulator input circuit. The output of the carrier oscillator is connected to the carrier input portion of the modulator; the effects are shown in the block diagram.

When two frequencies are combined or mixed in the modulator, both original frequencies can appear in the output together with several other frequency components. The direct result of the modulation process always is the production of upper and lower sidebands (see page 16). *Both* are invariably present at the modulator output. In some carrier systems, to be discussed later, only one sideband is transmitted. In these types, the unwanted sideband is removed by a filter inserted after the modulator.

Let us assume that the carrier frequency is 8,000 Hz. The combining or mixing of this carrier frequency with the voice signals produces these primary frequencies in the modulator output.

Original carrier oscillator .... 8,000 Hz
Carrier frequency plus voice signals:
  8,000 + (200 to 3,400) .... 8,200 to 11,400 Hz
Carrier frequency minus voice signals:
  8,000 − (200 to 3,400) .... 4,600 to 7,800 Hz

Each of these sidebands contains the voice signals or speech intelligence. Therefore, only one sideband, either the upper or lower, need be transmitted over the telephone line. The original 8,000-Hz frequency may or may not appear in the output of the modulator. It can be suppressed, if so desired, by the use of a balanced-bridge type modulator to be described later.

Demodulator Functions

The demodulator is an essential element of the receiving portion of each carrier channel. It operates in a manner similar to the modulator in the transmitting section. The demodulator functions may be considered as the reverse of those of the modulation process. Carrier signals reaching the demodulator are combined with the initial generated carrier frequency. The resultant demodulation process, to be described in detail later, will cause the original transmitted speech signals to be reconstructed.

Frequency Products of Demodulation

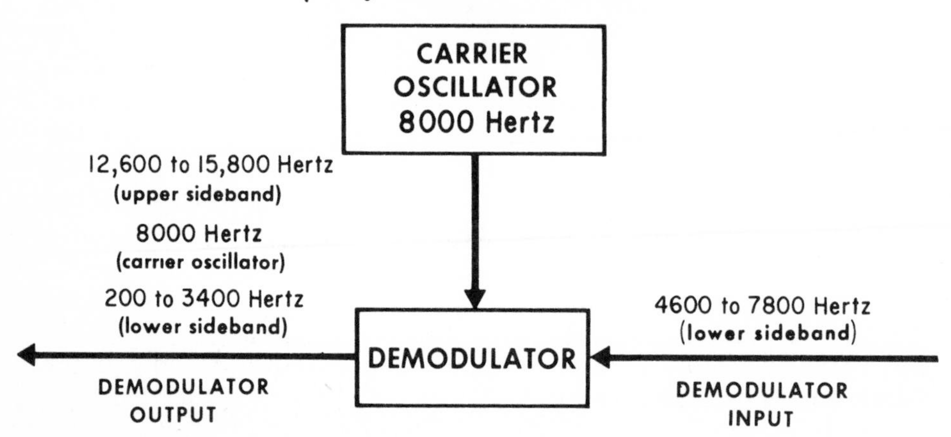

The diagram illustrates the carrier input signals to the demodulator and the resultant principal output frequencies. It is assumed in this example that the incoming carrier signals are in the range of 4,600 to 7,800 Hz. This is the lower sideband of the original transmitted carrier signals. The same 8,000-Hz carrier oscillator used for the modulation process is also employed in this demodulator. The main frequency products contained in the output of the demodulator are indicated in the diagram. The carrier oscillator frequency (8,000 Hz) is suppressed in this demodulator as it was in the modulator previously described.

This particular demodulation process, called *carrier reinsertion*, is required for carrier-suppressed types of systems. There are other carrier systems, such as the Western Electric N1 and N2, and the GTE Lenkurt types 47A/N1 and 47A/N2, in which the carrier frequencies are transmitted along with the sideband frequencies. The channel carrier frequencies need not be inserted at the receiving terminal for demodulation purposes. When carrier transmission is provided, the channel demodulation process may be considered similar to the action of the diode detector in a standard broadcast radio receiver.

Purposes of Filters

The modulation process will produce many frequency components in addition to the desired ones. As previously explained, it is necessary only to transmit either the upper or lower sideband over a carrier channel to provide speech intelligence. If all of the generated frequency components were transmitted, large frequency spacing between carrier channels would be required to prevent interference. This would reduce materially the number of carrier channels that could be sent over a telephone circuit. To suppress the undesired frequency components while permitting the free passage of the desired frequency range are the main purposes of the *filters* used in carrier channel equipment. Filters are devices that can discriminate between frequencies. They usually contain capacitive and inductive elements to form tuned circuits. Bandpass and low-pass filters are employed in carrier channels. Their design and operation will be described later. In general, the bandpass filter permits the passage of only the frequency range of a particular carrier channel. The low-pass filter is inserted in the demodulator's output to allow only the passage of the reconstructed voice-frequency signals.

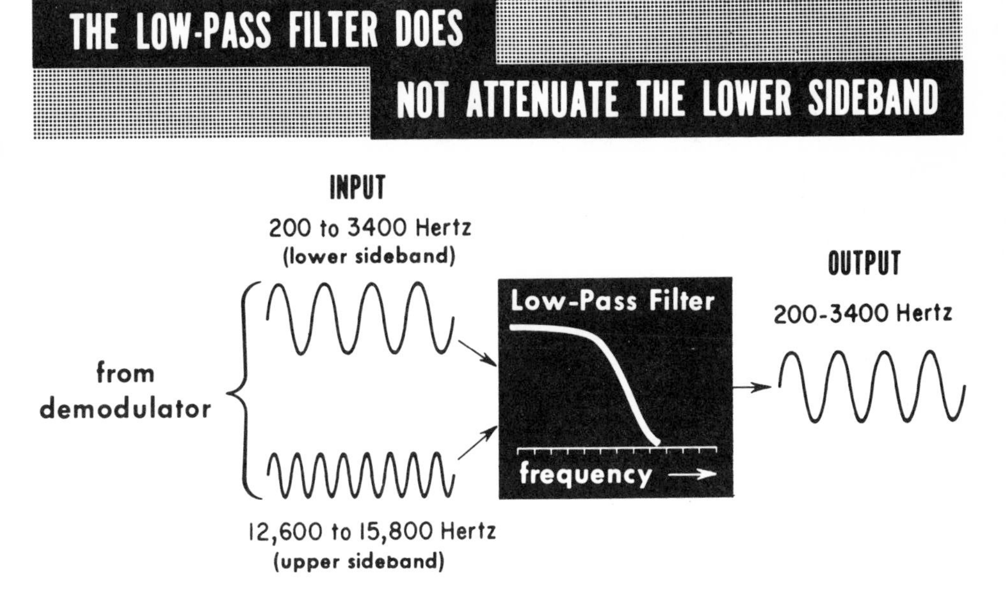

In the illustration, a low-pass filter is shown inserted in the output circuit of the demodulator. This allows the lower sideband which contains the reconstructed original speech to pass unaffected. The upper sideband or 12,600- to 15,800-Hz signals and other products of the demodulation process are greatly attenuated.

Transmitting and Receiving Amplifiers

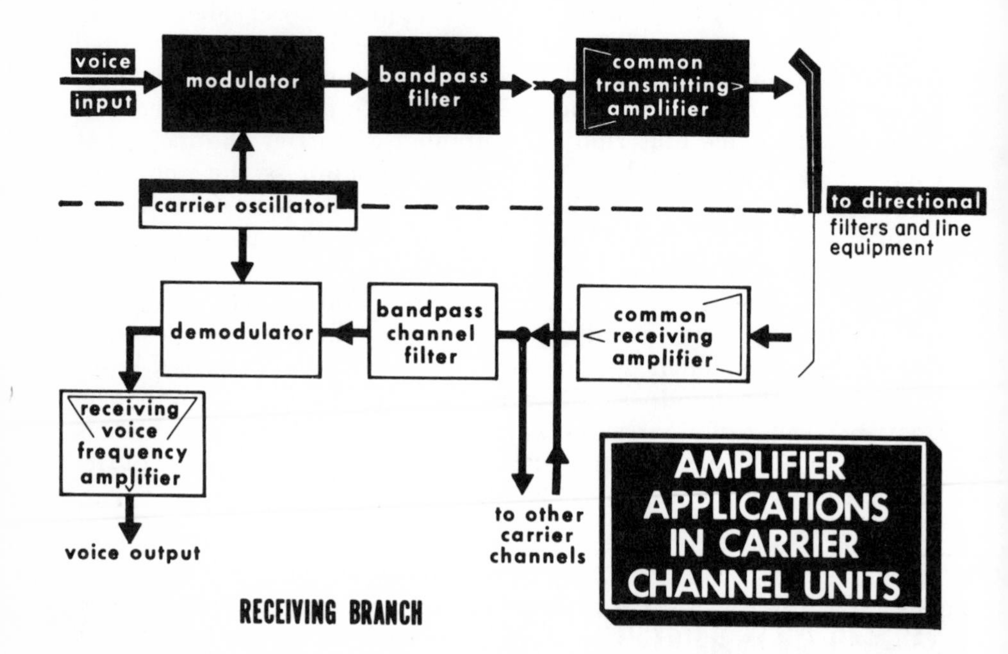

In addition to its individual carrier frequency oscillator, modulator, demodulator, and filters, each carrier channel may contain transmitting and receiving amplifiers. The channel *transmitting amplifier* boosts its carrier frequency level for transmission over the cable pair. The *receiving amplifier* in a channel usually follows the output of the demodulator to raise the voice signals to an adequate level. This is essential to take care of the transmission losses through balancing networks, hybrid coils, and other equipment connected to the toll trunk. In a carrier terminal, the various carrier channels may also feed their output carrier signals into a common *transmitting amplifier.* This will insure that all carrier channel signals are sent over the cable pair at equal strength. At the distant end of the line, a common *receiving amplifier* is usually provided in most types of carrier terminals. This amplifier increases the carrier signals to the proper level for ready passage through the various filters and the demodulation process.

The block diagram illustrates how amplifiers are used in carrier channel units. Because of long transmission lines, attenuation changes due to temperature or other conditions can seriously affect carrier signals on open-wire lines and in cables. To compensate for this, automatic gain regulation of the common receiving and transmitting amplifiers often is provided.

Amplifier Gain Regulation

In long open-wire circuits, the line attenuation often changes because of temperature effects, sleet, and other weather conditions. Aerial cable circuits similarly will be affected by day and night temperature variations. These attenuation deviations can adversely influence the performance of the carrier system. To compensate for these circuit attenuation variations, it is necessary to regulate the gain of the common receiving amplifier at carrier terminals and the gain of carrier repeaters. In suppressed carrier systems, amplifier regulation is obtained by an automatic gain control (AGC) circuit directed by a *pilot* frequency. In these systems, signals are transmitted only when one or more channels is being modulated. Therefore, it is necessary to provide one or more pilot frequencies on the circuit to control the AGC functions when no channel signals are present. Pilot frequencies are not necessary for this purpose in carrier transmitted type systems. The channel carriers are always present to furnish the required signal levels for controlling AGC operations.

To illustrate the principles of amplifier AGC operation better, reference is made in the block diagram to an old type of carrier system in which the gain of the amplifier is fixed. An electronic variable attenuator controls the amplifier's output. Let us see how it functions in a typical two-wire carrier terminal. The signals received from the distant West carrier terminal include the channel carrier frequencies and the 20-kHz pilot frequency. All of these signals pass through the receiving directional filter, line equalizer, and the variable attenuator of the AGC circuit to the common receiving amplifier. There are two paths at this amplifier's output: one to the channel bandpass filters and the other to the pilot frequency 20-kHz filter. The channel carrier frequencies are accepted by their respective channel bandpass filters but are rejected by the 20-kHz pilot filter. The 20-kHz pilot frequency,

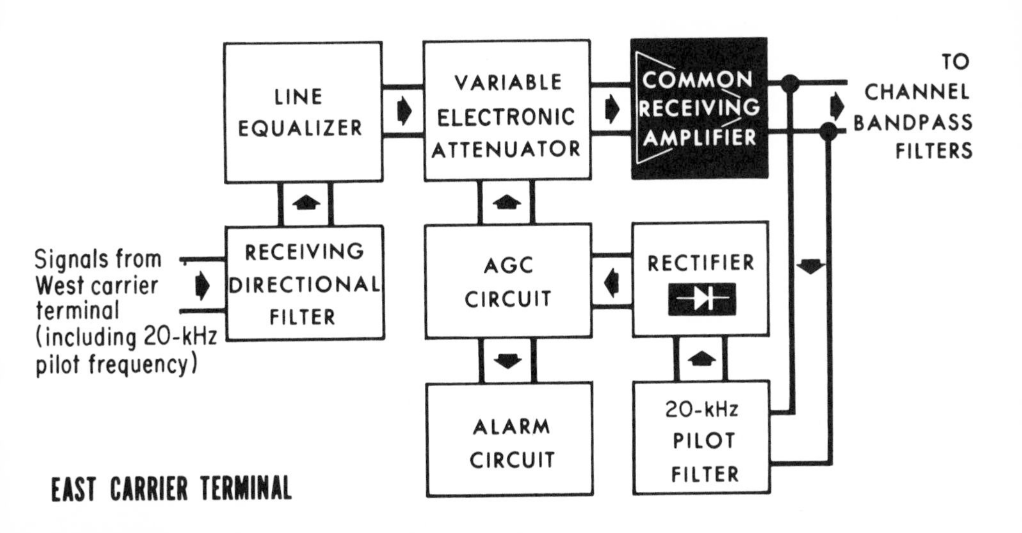

Amplifier Gain Regulation (cont.)

however, can go through the pilot filter to the rectifier circuit which usually comprises semiconductor devices to control the AGC circuit. In normal operation, the 20-kHz pilot frequency is at a predetermined level and the AGC circuit is inoperative. An increase in the attenuation of the wire line will cause a reduction in the level of the pilot frequency. This, in turn, will cause the AGC circuit to direct the electronic variable attenuator to increase the gain of the common receiving amplifier. The increased output of the amplifier will restore the pilot frequency to its proper level and release the AGC circuit. Also, a decrease in the line's attenuation will make the electronic variable attenuator reduce the output of the common receiving amplifier until the pilot frequency is restored to its predetermined level.

The regulation of line receiving amplifiers and repeaters in present carrier systems is achieved electronically by the feed-back technique. A portion of the amplified output is sent back to the input stage to control the bias and gain of that stage. To prevent rapid gain changes, the AGC feed-back path includes a *thermistor,* or thermal resistor. This device varies its resistance *inversely* with the amount of current through it. The current flow causes heating of the thermistor, and the temperature determines its resistance. The higher the temperature, the lower will be its resistance and vice-versa. A time lag is designed into the thermistor so that resistance changes will be gradual in order to prevent abrupt changes in amplifier gain.

An alarm circuit is usually associated with the AGC circuit. Large changes in pilot frequency levels due to line breakage, short circuits, or other reasons will actuate the alarm circuit to notify maintenance personnel.

Questions and Problems

1. What range of frequencies can be transmitted effectively over open-wire and cable telephone circuits?
2. What basic multiplex method is usually employed by carrier systems?
3. What are carrier signals? How are they produced?
4. Why is frequency coordination important in carrier systems?
5. What means may be employed to reduce crosstalk interference?
6. Name the basic components of a carrier channel.
7. What is produced by the modulating process?
8. What is the end result of demodulation?
9. What main functions are performed by filters in carrier systems?
10. What method is usually utilized to regulate the gain of amplifiers in suppressed carrier systems?

Modulation Devices

The mixing or combining of frequencies in carrier equipment is the function of the *modulator* and *demodulator*. Solid-state devices such as transistors, diodes, and integrated circuits (IC) are used to perform the mixing process and amplification. The early carrier systems used vacuum tubes in the modulators and demodulators. With the development of satisfactory semiconductor devices, the tubes were soon replaced by them. The first such solid-state devices employed were the copper-oxide rectifier and the *varistor*. The varistor is a two-electrode semiconductor device having a voltage-dependent nonlinear resistance. That is, its resistance changes with the magnitude or direction of the applied signal voltage.

The drawing shows the standard symbol for solid-state diodes. This category includes the Varistor and the copper-oxide rectifier. The several basic schematics serve to illustrate the operation of these solid-state devices. The head of the arrow in the figures represents the direction of electronic (plus to minus) current flow. Using the electron flow concept, the vertical line of the diode symbol is considered to be the cathode.

Manufacturers usually mark the diode's cathode with the "+" sign as an indication that it should be connected to the positive potential in a power supply circuit. The path of electron flow opposite to the arrowhead is termed the *forward* direction. By convention, the opposite path is called *reverse*. In the forward direction, the diode resistance is very low; in the reverse direction, its resistance is very high.

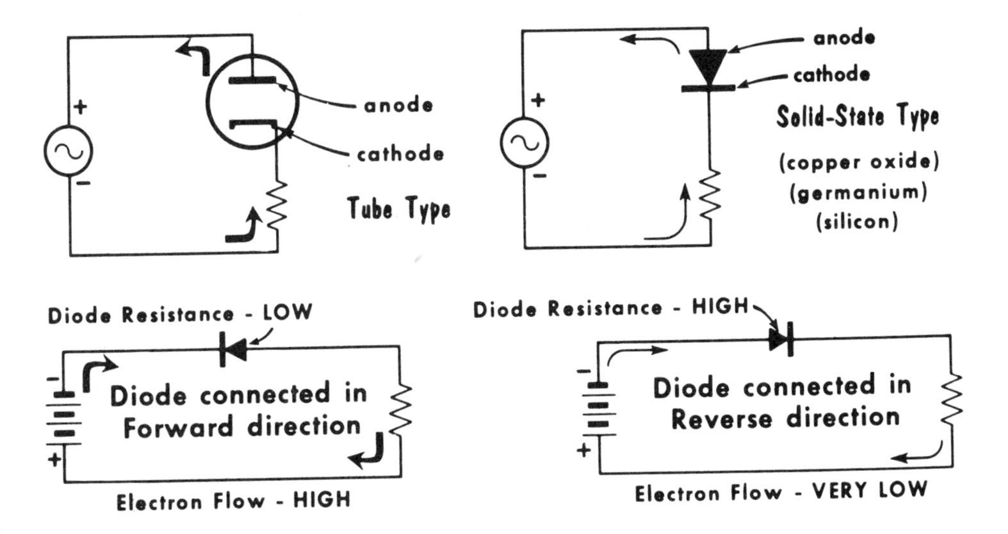

Semiconductor Characteristics

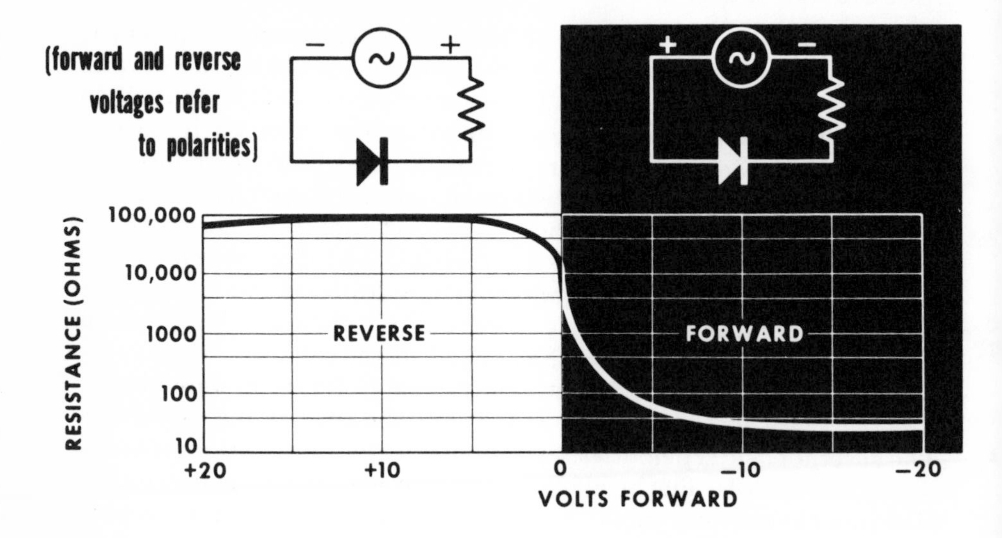

RESISTANCE CHARACTERISTIC OF A CRYSTAL DIODE

Copper-oxide rectifiers had been in use in carrier telephone systems for many years. They were more economical and stable than vacuum tubes for providing the modulation and demodulation operations. In recent years, germanium and silicon diodes have largely replaced the copper-oxide rectifier. These diodes have a much higher resistance than copper-oxide devices. Therefore, they can perform the rectifying and mixing operations more efficiently.

The graph illustrates the resistance variation characteristics of typical rectifiers used in modulators and demodulators. If the polarity of the applied signal voltage should be in the reverse direction through the diode, for example, an extremely large resistance to current flow would result. This high resistance could be considered an open circuit. If the voltage polarity were now reversed, the resistance presented by the rectifier would decrease to a very low value of around 30 ohms. This would be equivalent to a short circuit. Therefore, a semiconductor diode can be considered to have the characteristics of a voltage-controlled switch. In other words, the magnitude and polarity of the applied signal voltage will open or close the switch. By applying this technique to a carrier terminal, we can see that the carrier signals will switch the diode units in the modulator between almost zero and a very high resistance value, at the carrier frequency rate. This switching operation is the heart of the modulation and demodulation process.

Balanced-Bridge Modulator

Two types of modulators and demodulators (modems) are commonly used in carrier systems; *balanced-bridge* or *shunt-type*, and *lattice* or *ring-type*. The balanced-bridge modulator will be discussed first. Since demodulators operate in the same manner as modulators, this discussion applies equally to both devices. The carrier, which transmits no voice intelligence, normally is suppressed in the modulator circuit, with the exception of carrier transmitted systems, which are analyzed on pages 75 and 113.

A simplified schematic of the balanced-bridge or shunt-type modulator is shown. The four diodes, arranged in a Wheatstone bridge configuration, alternately pass and short circuit the input voice signal at a rate equal to the carrier frequency. Let us apply an alternating current or a carrier voltage to terminals 2 and 4 so that, during one-half cycle, terminal 2 is negative and terminal 4 is positive. No current will flow from terminals 2 to 4 because, with this polarity condition, the diodes are biased in the reverse direction and offer a very high resistance. Consequently, an applied voice signal to terminals 1 and 3 would not be affected.

In the next half cycle the polarity will be reversed. Terminal 2 will become positive and terminal 4 will become negative. Thus, the four diodes will be biased in the forward direction, presenting very little resistance, and the carrier current will flow from terminal 4 to terminal 2; one half through terminal 1 and one half through terminal 3. The voice input signal, in this situation, will be short circuited.

The modulator output will have the form of the wave shown on page 74. The carrier frequency, however, will not appear in the output because terminals 1 and 3 are points of equal potential with respect to the carrier source. It should be noted that this condition depends on the accuracy with which the four diodes are matched to provide equal balance of each leg of the bridge. For effective modulation, it is necessary that the voice signals adequately control the diodes.

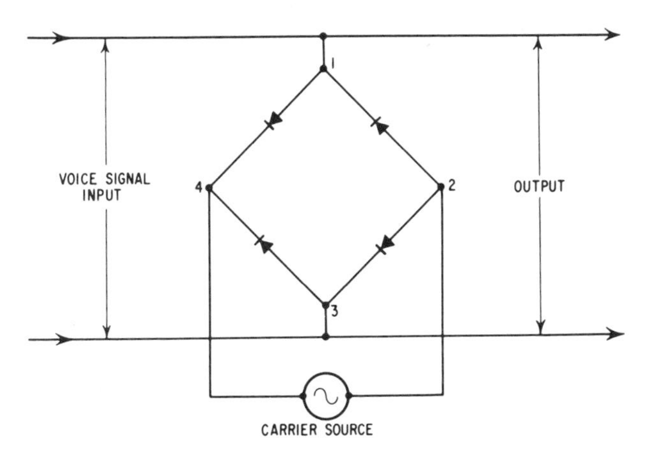

Voltage Switching in Balance-Bridge Modulator

We shall now further study the aforementioned principles to better understand the operation of the balanced-bridge modulator. First, it is desirable that we know how the carrier voltage can control the switching action of the semiconductor diodes or rectifier elements. Figure A is a simplified circuit of a balanced-bridge modulator. A source of carrier frequency voltage only is applied to the rectifier units.

Assume that the applied carrier frequency has the polarity indicated during one part of the cycle. There will be a negative potential at terminal 2 and a positive potential at terminal 4. Current will not flow in the rectifying units under this condition. This means, as previously explained, that the diodes have a very large resistance at this instant. This is tantamount to an open circuit as per the equivalent circuit shown in Fig. B. In the next part of the cycle, the polarity of the carrier frequency is reversed (Fig. C).

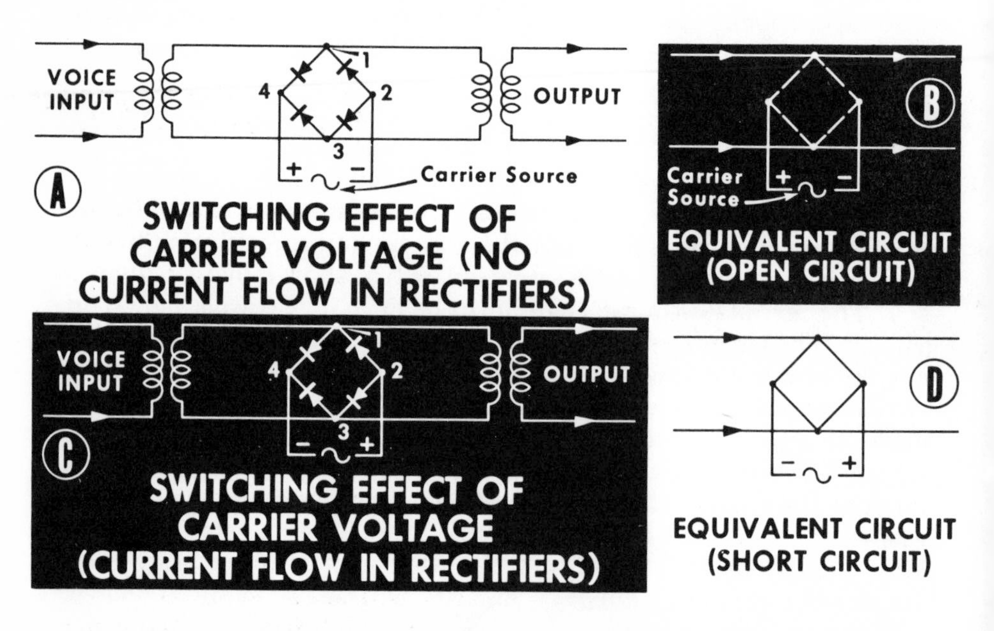

Now, terminal 4 is negative and 2 has a positive potential. This is the reverse of the previous condition of the rectifier elements which causes them to have a very small value of resistance. In essence, they serve as a short-circuit across the voice signal path (Fig. D). This process is repeated at the frequency of the applied carrier voltage. For example, if the carrier frequency is 10,600 Hz there will be 10,600 open-circuits and 10,600 short circuits per second across the voice path. Thus, the carrier frequency voltage, in effect, will open and close the diode switches a total of 21,200 times each second of applied carrier voltage.

Operation of Balanced-Bridge Modulator

(Arrows indicate voice frequency currents)

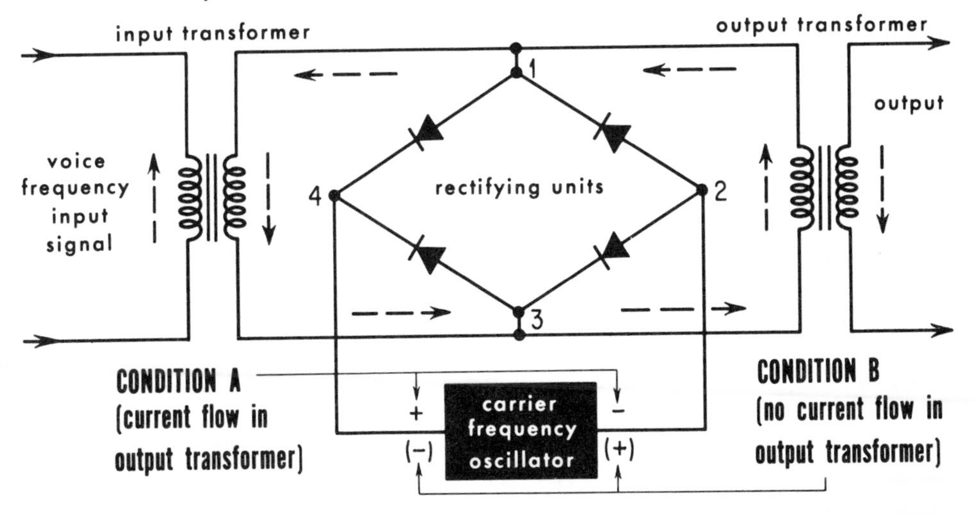

BALANCED-BRIDGE MODULATOR OPERATIONS

The operation of a balanced-bridge modulator will now be followed in detail. Referring to the diagram, we see a simplified schematic circuit of a balanced-bridge modulator including a voice-frequency source, as well as the carrier frequency.

Let us assume (condition A) that the instantaneous carrier frequency polarity is such that terminals 2 and 4 are negative and positive, respectively. The rectifying units will thus be momentarily reverse biased and appear to be open circuited. At the same time, it is assumed that the voice signal has a voltage polarity so that the current flows as indicated by the arrows. Voice currents will not go through the rectifying units because they are effectively open circuited. At a later instant, the polarity of the voice frequency voltage will be reversed but the carrier frequency's polarity will be the same as shown. Since the rectifying units still will appear to be open circuited, voice currents will not be affected and will continue to flow through the output transformer, although in an opposite direction.

A reversal of the carrier frequency polarity (condition B), however, will result in no flow of voice frequency current in the output transformer. For instance, if terminals 2 and 4 were positive and negative, respectively, the rectifying elements will be forward biased and appear to be short circuited. Therefore, there will be a short circuit across the input transformer and voice currents cannot pass to the output transformer. This will take place regardless of the polarity of the voice signal voltage.

Balanced-Bridge Modulator Output Waveform

We shall now examine in more detail the resultant output of the balanced-bridge modulator. For this purpose it is assumed that the carrier-frequency voltage has a positive polarity. The voice-frequency currents can pass through the modulator's output circuit during this positive portion of the voltage cycle. When the carrier voltage goes negative, the voice-frequency path becomes short circuited by the switching effect of the rectifiers. There will be no modulator output voltage during this period.

OUTPUT OF THE BALANCED-BRIDGE MODULATOR

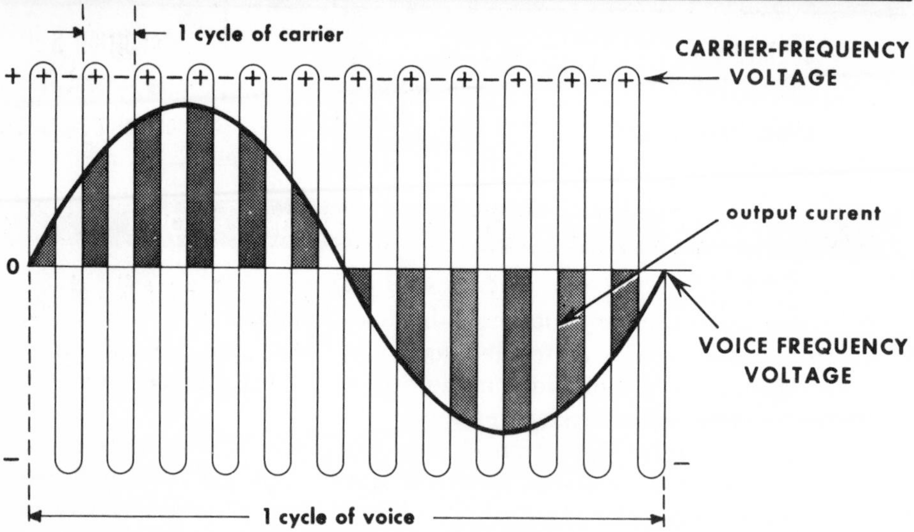

Shaded area represents the amount and polarity of the voice frequency current in the output of the balanced-bridge modulator.

The graph shows the output waveform during one cycle of the voice-frequency voltage, which includes numerous carrier-frequency voltage cycles. The positive and negative signs indicate the polarity of the carrier voltage. As explained above, the voice-frequency current can flow only in the output circuit during the positive cycles of the carrier voltage. The polarity and intensity of the voice frequency current in the modulator's output are represented by the shaded area. The waveform of the modulator's output is made up of many different frequencies. For wire carrier use, we are concerned only with the original voice-frequency, upper sideband, and lower sideband. The original carrier-frequency does not appear in the modulator's output. It is shunted out by the switching effect of the rectifier elements, as previously described.

Balanced-Bridge Modulator: Carrier Transmitted Systems

The major output frequencies of the balanced-bridge modulator can be classified as voice (V); carrier plus voice (C + V) termed the upper sideband; and carrier minus voice (C − V) termed the lower sideband. Other frequencies, harmonically related to the carrier and voice frequencies, are also present. These unwanted frequencies are low in amplitude and are later removed by the filters in the carrier terminal equipment.

When direct current is passed through opposite sides of the balanced-bridge modulator, we can intentionally unbalance the bridge circuit. This will cause the carrier frequency to be present in the output of the carrier channel along with the voice signals and the upper and lower sidebands. This arrangement is illustrated in the schematic diagram. It is used in carrier transmitted systems such as the Western Electric N-1, N2, and the GTE Lenkurt types 47A/N1 and 47A/N2. Balanced-bridge modulators, therefore, may be adapted for either *carrier suppressed* or *carrier transmitted* types of systems. This option, incidentally, is not possible with the lattice type modulator next to be described. In essence, the balanced-bridge modulator has its chief application in the carrier channel element of the system.

BALANCED-BRIDGE MODULATOR FOR CARRIER TRANSMITTED SYSTEM

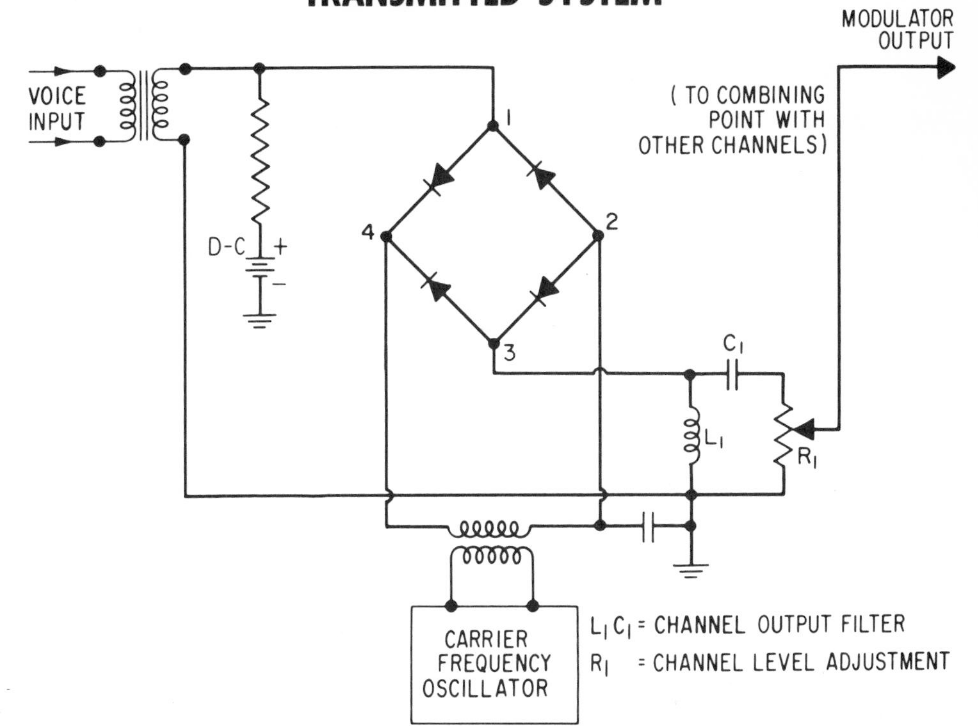

Lattice Modulator Bridge Circuit

The *lattice modulator* circuit is another arrangement of varistors or semiconductor diode rectifiers to perform the modulation and demodulation operations in a carrier terminal. There are two methods of arranging the rectifier units in a lattice modulator. The lattice modulator bridge circuit is shown below.

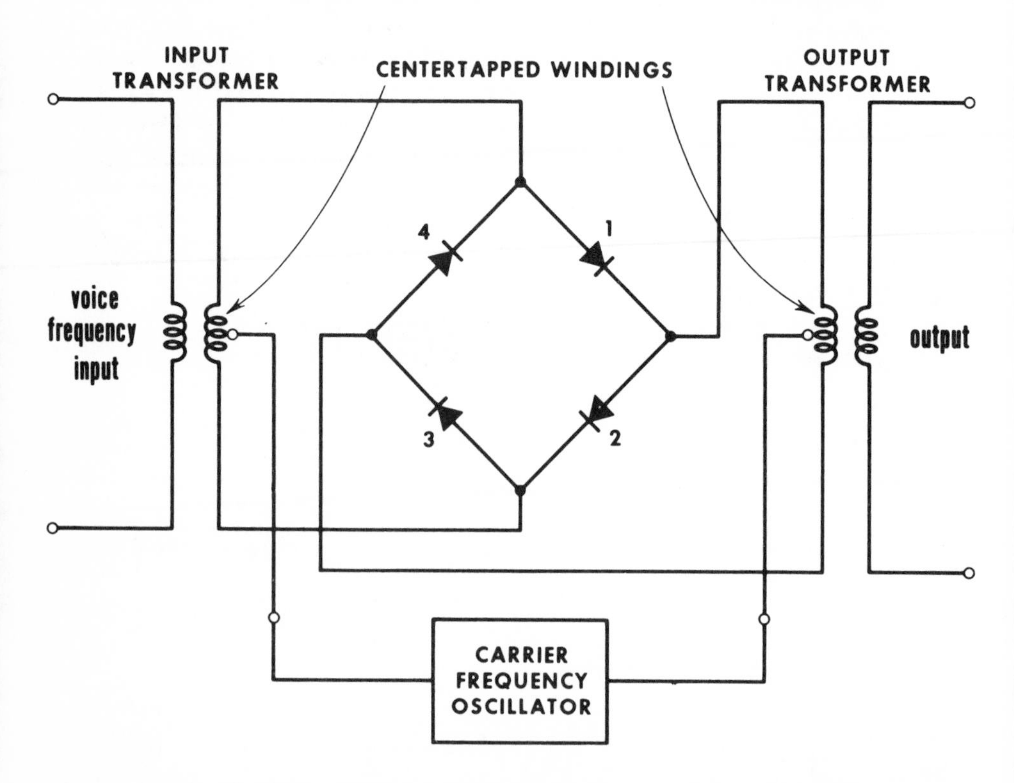

Note that the rectifier elements are arranged in the well-known Wheatstone bridge circuit. It is similar to that of the balanced-bridge modulator previously discussed. The connections for the carrier frequency source, however, are quite different. Note that the input and output transformers have centertapped windings that connect to the carrier frequency source. This permits the carrier frequency currents to flow in opposite directions through these transformers, as we shall subsequently learn. This will make it possible for the carrier voltages in the transformers to balance or cancel out each other. Therefore, no carrier frequency voltage will be induced into the output circuit of the lattice modulator.

Lattice Type Modulator Circuit

The more common type of lattice modulator circuit is shown below. In this arrangement, the varistors or rectifiers form a symmetrical lattice or crossover network. This type of circuit is used extensively in present-day carrier systems.

The carrier frequency voltage is applied at the centertaps of the transformer windings as in the lattice modulator bridge circuits. The voice frequency signals are applied to the lattice arrangement by transformer coupling. The rectifier units operate as switches under control of the carrier frequency voltage. They reverse the direction of the voice frequency current in the output of the modulator at the rate of the carrier frequency. It is essential that the carrier frequency voltage be much higher than the voice frequency voltage for proper operation. In general, the lattice type modulator combines the voice and carrier frequencies in such a manner that the modulator's output will contain only the upper and lower sidebands. The carrier frequency and voice signals are suppressed. This represents a considerable advantage over the balanced-bridge modulator circuit.

Common Lattice Modulator Circuit

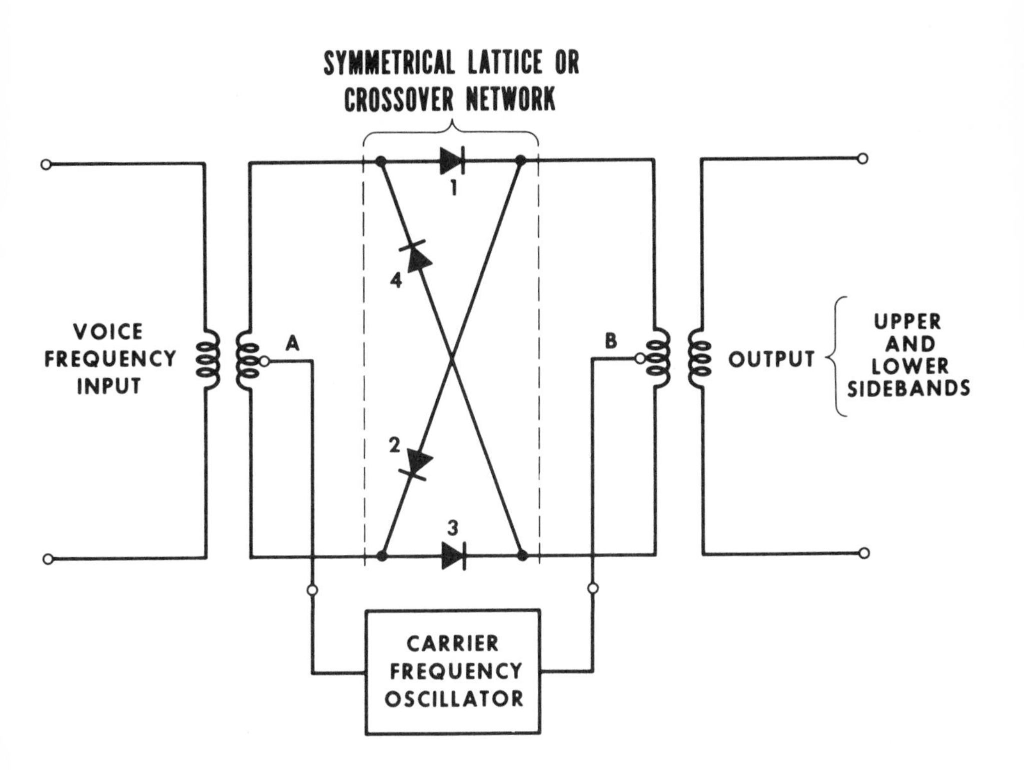

Carrier-Controlled Switching (Open Circuit)

It is desirable that we know how the carrier voltage controls the switching in the lattice modulator. This is essential to our better understanding of the operation of the lattice type modulator.

SIMPLIFIED SCHEMATIC OF LATTICE MODULATOR

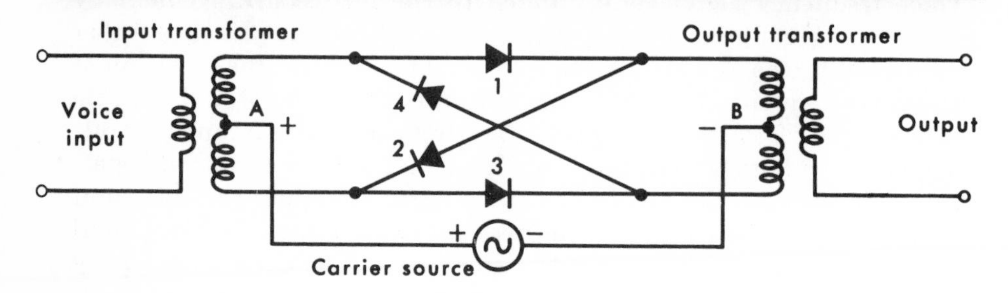

A simplified schematic of a lattice modulator circuit is shown above. A source of carrier frequency voltage is connected to the centertaps of the input and output transformers. The carrier voltage is assumed to have the polarity as indicated. With this polarity, rectifier units 1 and 3 are short circuited because current flows through them. Rectifiers 2 and 4, on the other hand, will be open circuited. They have very high resistance to the designated voltage polarity.

RESULTANT PATH OF CARRIER CURRENTS

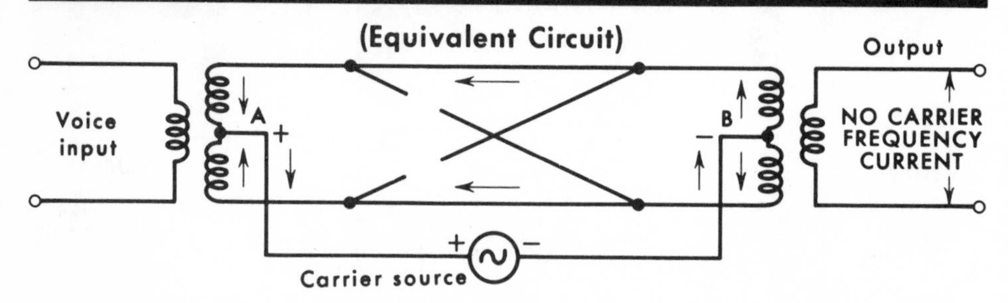

The resultant path for the carrier currents is illustrated above. Note that the carrier current divides equally at the transformer centertaps A and B, respectively. The currents flow in opposite directions as shown by the arrows. Since there are equal currents flowing in opposite directions through the centertapped windings of the transformers, no voltage is induced in the other windings of the transformers. Therefore, no carrier frequency current appears in the transformer's input and output circuits.

Carrier-Controlled Switching (Short Circuit)

When the carrier frequency potential reverses, a different condition exists. Rectifiers 2 and 4 are now short circuited due to the carrier current flowing through them. Rectifiers 1 and 3 are open circuited under this condition.

SIMPLIFIED SCHEMATIC OF LATTICE MODULATOR

Voice input
A
B
1
2
3
4
Output
Carrier source

The path of the carrier frequency current is changed as a result of the above switching. The new equivalent circuit arrangement is shown below. In this case, note that the transformer windings are still connected in parallel but in the reverse order of that appearing on the preceding page.

RESULTANT PATH OF CARRIER CURRENTS

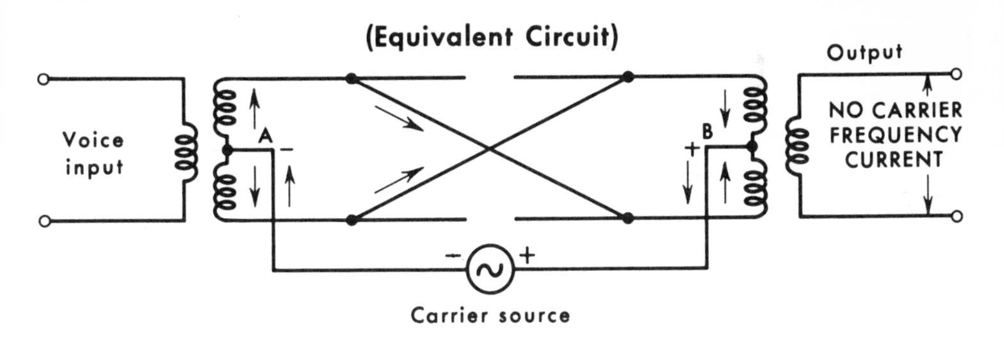

The carrier frequency currents again will divide equally at the centertapped windings of the transformers. They will flow in opposite directions at the centertaps marked A and B. Therefore, no voltage can be induced in the other windings of the transformers which connect to the input and output circuits, respectively. The reversal of the carrier polarity has changed or switched the closed-circuit path between the two centertapped windings. In effect, we can say that the polarities in the output circuit have been reversed with respect to the voice frequency input polarities at the carrier frequency rate.

Switching of Voice Currents in Lattice Modulator

Let us now apply the foregoing principles to the operation of a lattice type modulator. In the two schematic drawings, the arrows show the paths taken by the voice frequency currents under the two indicated polarities of the carrier frequency source.

Operation of the Lattice Type Modulator

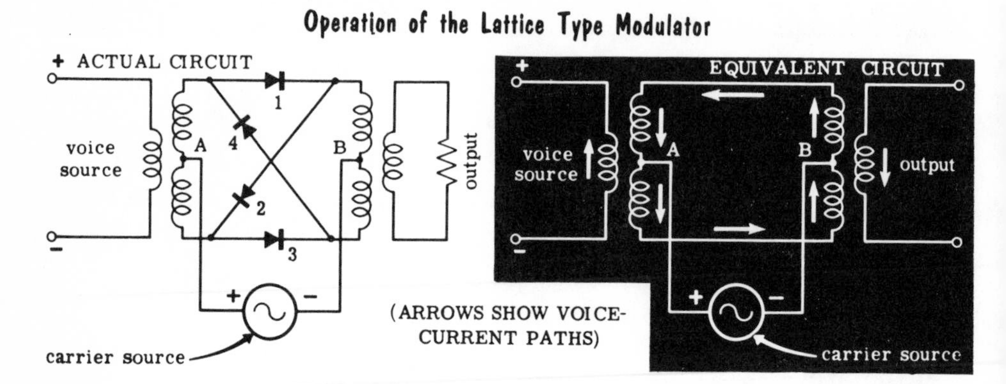

In the above diagram, the polarity of the carrier frequency at centertaps A and B causes rectifiers 1 and 3 to conduct. Rectifiers 2 and 4, therefore, act as an open circuit. The resultant flow of the voice frequency currents is as shown by the arrows.

Carrier Source Polarity Is Reversed

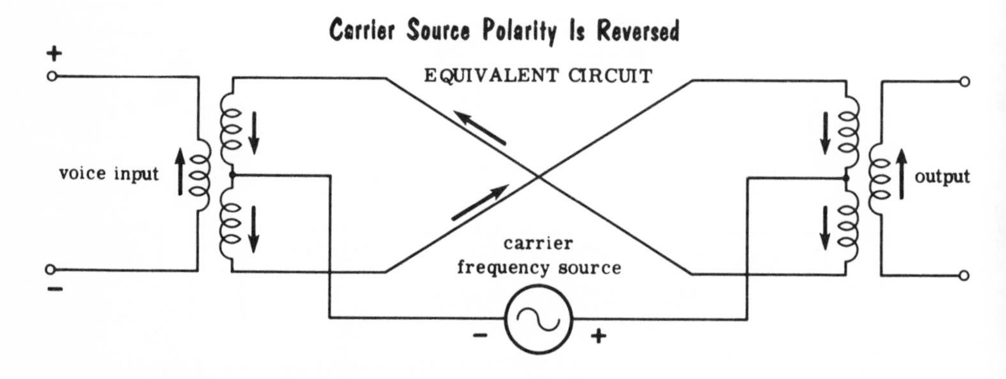

In the above illustration, the polarity of the carrier source has been reversed. Rectifiers 2 and 4 now conduct, and 1 and 3 are open circuited. The path between the two transformer windings are reversed, as shown. This will also reverse the direction of the flow of the voice frequency currents through the output transformer. The direction of the voice frequency currents through the input transformer, however, has not been changed.

Effect of Carrier Switching in Lattice Modulator

The frequency of the carrier source, as we have learned, must be higher than that of the voice frequency or speech signal. The carrier oscillator, for example, may have a frequency in the range of 6,000 to 150,000 Hz (6 kHz to 150 kHz). Therefore, the polarity of the carrier frequency signal changes many times before the voice frequency changes once. The resultant current traveling through the output transformer is reversed at the carrier frequency rate. As an example, let the carrier frequency be 13,600 Hz. We will first consider a particular speech signal having a frequency of 400 Hz. During one cycle of the 400-Hz speech signal, there will be 34 cycles of 13,600-Hz carrier frequency. This also means that the 400-Hz speech signal will be switched 34 times during each of its positive polarity half-cycles. Similarly, there will be 34 reversals during each of its negative half-cycles.

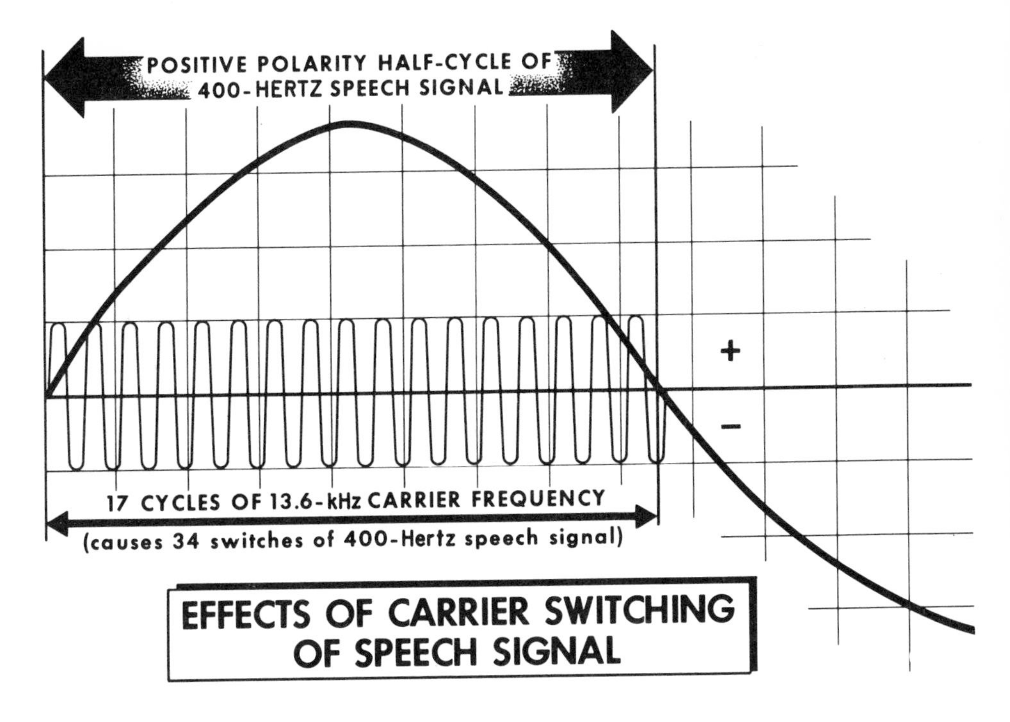

EFFECTS OF CARRIER SWITCHING OF SPEECH SIGNAL

The current continues to flow in the output of the modulator irrespective of the changing carrier frequency polarity. The carrier frequency polarity controls the direction of the output current for a given polarity of the voice frequency or speech input signal. The amount of the output current depends upon the instantaneous value of the voice frequency voltage. Therefore, as the direction of the output current changes at the carrier frequency rate, its value varies in accordance with the voice frequency voltage.

Lattice Modulator Output Waveform

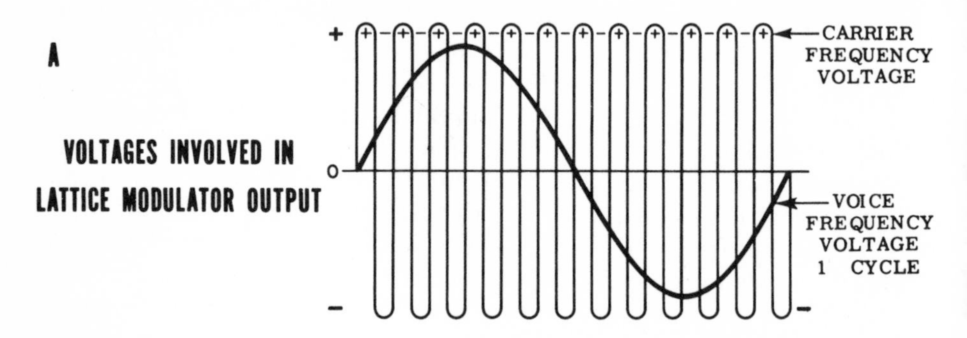

We shall now develop the waveform of the lattice modulator's output. First, let us look at diagram A. It was also used in the development of the waveform for the balanced-bridge modulator on page 74.

Assume that the same one voice frequency cycle and the many carrier frequency cycles are applied to the lattice modulator. Starting with the positive half-cycle of the carrier source, the speech signal starts to increase from the zero axis. The output current likewise increases. During the next negative half-cycle of the carrier frequency, the output current's direction changes. Its value, however, continues to increase. As this process continues, the envelope of the lattice modulator's output current follows the curve of the voice frequency signal.

Diagram B shows the output waveform of the lattice modulator. Note that the current envelope follows the curve of the speech signal. The current, therefore, is always flowing in the output circuit regardless of the polarity of the carrier source. This is a different condition than that existing in the balanced-bridge modulator. The result is that there is more power in the output of a lattice modulator than in the balanced-bridge type. In addition, the lattice modulator's output contains only the upper and lower sidebands. The carrier and voice frequency signals are effectively suppressed by the action of the varistors or rectifier units, and the balanced windings of the input and output transformers.

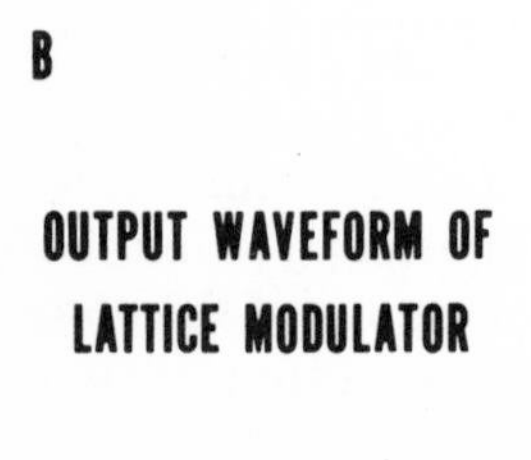

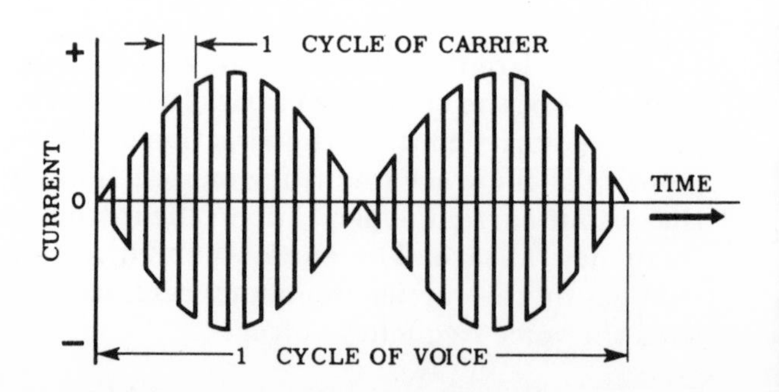

Multiplex and Modulation Types

Two principal types of multiplex, the technique of sending many voice channels over the same wire or radio transmission facility, are used in carrier systems: *frequency division (FDM)* and *time division (TDM) multiplex.* Frequency division (the basis of almost all systems until about 1961) uses two basic modulation processes: *amplitude modulation (AM)* and *(FM) frequency* or *phase modulation.*

Amplitude modulation is the most common type in cable carrier systems for both short and long haul transmissions. It permits one frequency to control the instant amplitude of another, usually the carrier frequency. As an illustration, the diagram shows a 12,000-Hz carrier frequency amplitude modulated by a pure 1,000-Hz tone. The relative amplitudes of the carrier, the modulating tone, and the resultant upper and lower sideband products are plotted. The carrier frequency and sidebands have much higher amplitudes than the original tone signal. Voice signals will cause similar variations in carrier amplitude.

Frequency and phase modulation carrier systems have been employed mainly on subscriber and some interoffice trunk cables, but they are also used on open-wire lines, either alone or frequency coordinated with other carrier systems. The big advantage of FM and phase modulation is noise reduction ability. This gain is offset, however, by the large bandwidth needed—approximately 15 kHz for each voice channel. This wide bandwidth requirement reduces the distance over which such carrier systems can be operated economically. We will confine our carrier discussions to the amplitude modulation mode, al-

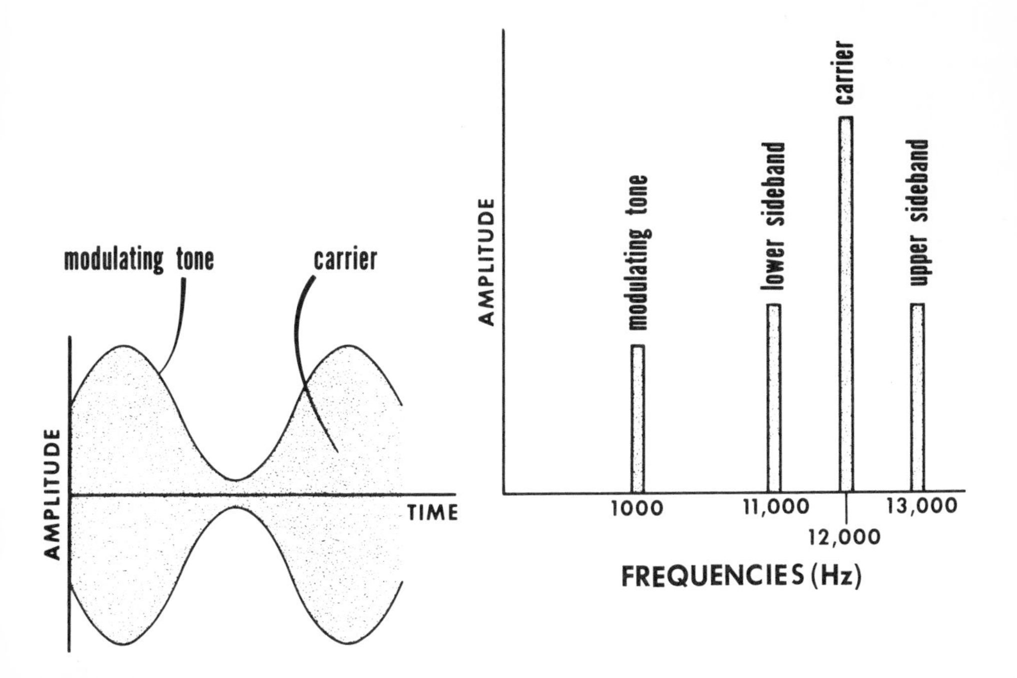

Multiplex and Modulation Types (cont.)

though the principles of frequency translation also apply to the FM and phase modulation types.

Improved engineering concepts concerning time division multiplex (TDM) and better solid-state devices have resulted in the development of *Pulse Code Modulation (PCM)* carrier systems. PCM carrier systems, as typified by the Western Electric T-1, utilize the time division sampling techniques of *Pulse Amplitude Modulation (PAM)* plus the considerable improvement in signal-to-noise performance made possible by pulse code modulation. This class of carrier was initially designed to expand the trunking facilities between central offices in and around large cities and their suburban areas. Improved techniques have made PCM carrier economically attractive for both short and long haul applications over cable and microwave radio facilities. Voice signals in the conventional FDM carrier systems are translated into separate frequency bands by the modulation method. In the typical PCM carrier, the voice signals are multiplexed by being sampled in time as pulses. We will study PCM further in a subsequent section.

Questions and Problems

1. What main function is performed by the modulator?
2. What electronic devices commonly are employed in modulators?
3. Name the general types of modulators and demodulators.
4. How many rectifying elements are required in each type of modulator and demodulator?
5. What principal frequencies appear in the output of a balanced-bridge modulator?
6. Can the balanced-bridge modulator be utilized for carrier transmitted systems? Explain.
7. What are the two basic circuits in a lattice modulator?
8. What prime frequencies appear in the output of a lattice modulator?
9. How does the power output of the lattice modulator compare with the balanced-bridge type?
10. Assume the carrier voltage in a lattice modulator has a frequency of 24,400 Hz. How many times will each negative half-cycle of a 400-Hz tone signal be switched?

Function of Filters

In our study of the *FDM Modulation and Demodulation Process,* we learned that the output waveform of the modulator (and also that of the demodulator) always contained the upper and lower sideband frequencies. They are the desired frequencies resulting from the modulation or demodulation process. Unwanted frequencies also are present in the output, together with these sidebands. To eliminate or suppress these undesired frequency components, use is made of certain circuit arrangements called *electrical filters.* It is the purpose of electrical filters, more commonly termed *filters,* to permit only the desired band of frequencies to pass. All other frequencies are eliminated or effectively suppressed.

The diagram shows a simple application of an electrical filter. The capacitor in this circuit is used to stop or block the flow of direct current from the battery to the loudspeaker. The audio signals, however, can be passed by the capacitor to the loudspeaker. Therefore, when the switch is closed, the lamp will light but the loudspeaker will not be affected by the direct current. Similarly, the audio-frequency currents are readily passed by the capacitor to activate the loudspeaker. Therefore, by permitting the passage of alternating current but not direct current, the capacitor serves as a filter.

Application of an Electrical Filter

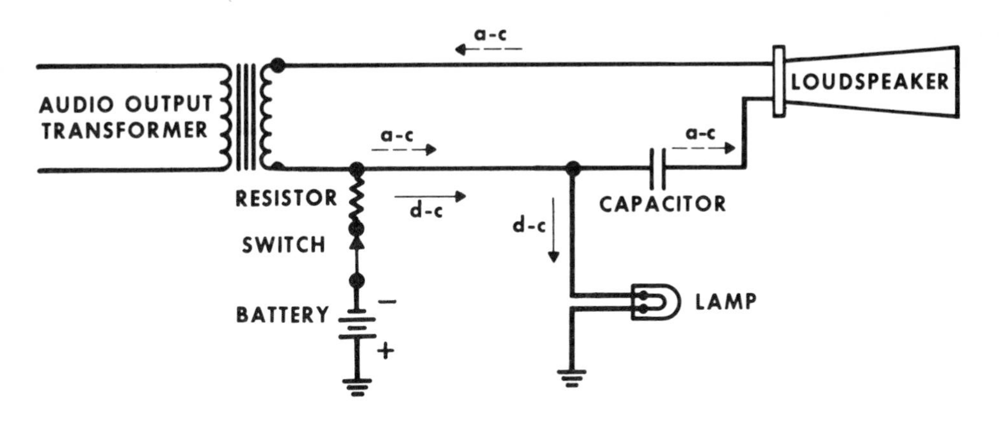

Capacitor presents 'open circuit' to the flow of d-c.

A-C can pass through capacitor

Filter Elements

To comprehend how an electrical filter functions, it is necessary that we know something about its composition. If we should look inside a simple filter as used in carrier systems, we would find that it consists of two main electrical elements—inductors and capacitors. These elements can be connected, for example, to form a tuned circuit which is resonant to one frequency or to a particular band of frequencies.

POWER SUPPLY FILTER ELEMENTS

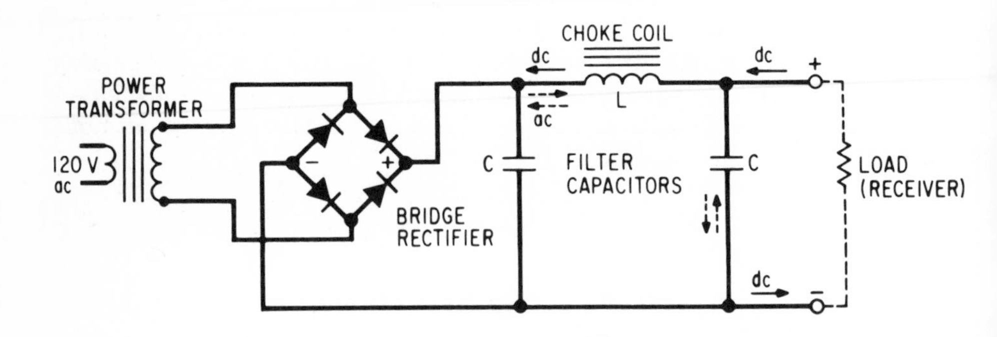

The simplified schematic diagram shows a power supply such as used in a broadcast radio receiver. A large inductor, in this case a choke coil, and two filter capacitors are connected to form a filter. The combination of the choke coil and capacitors form a tuned circuit. This circuit arrangement has a very low impedance to the flow of the low-frequency (60 Hz and its harmonics) alternating current from the rectifier's output. This provides a low resistance path to shunt the ac voltage present in the output of the rectifier. As a result, the ac voltage components are effectively short-circuited and only the direct-current (dc) output from the rectifier is permitted to flow to the radio receiver circuit. It follows, therefore, that two elements, inductors and capacitors, are required to form a filter that will respond to specific alternating currents, such as voice and carrier frequencies.

Inductive and Capacitive Reactances

Simplified Schematic of Repeating Coil

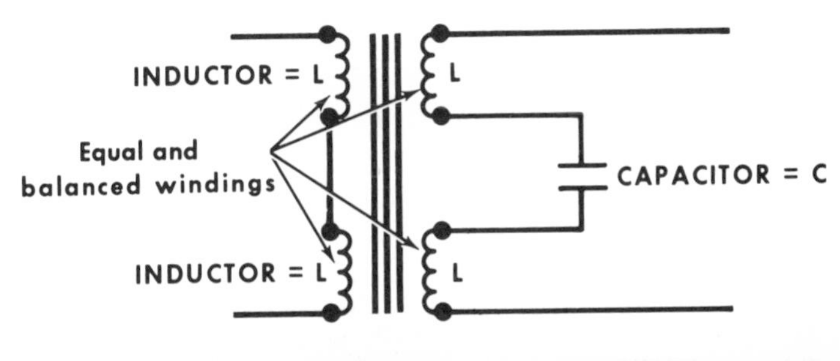

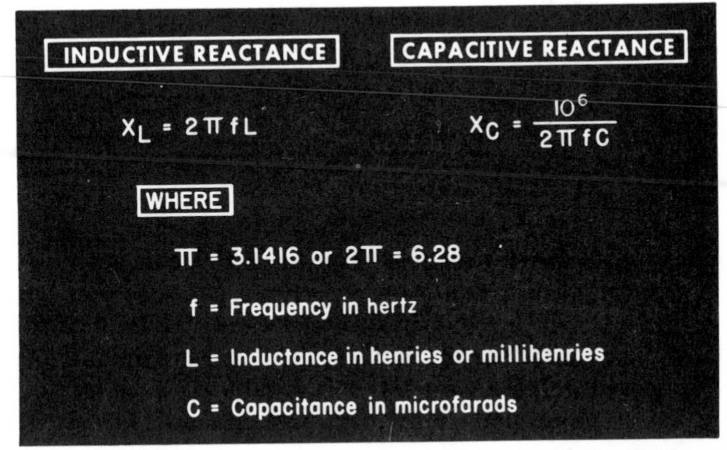

The proper understanding of filter principles and the design of filters will probably necessitate a brief review of certain basic electrical properties and formulas. For example, a coil of wire wound over an iron core has a certain inductance value.

The diagram shows a simplified drawing of a repeating coil with a capacitor connected between the two balanced windings. The basic formulas for calculating the inductive and capacitive reactances are also indicated. From the formulas, note that the inductive reactance, X_L, increases as the frequency increases. This causes a greater opposition to the flow of alternating current such as speech or voice frequency signals than it does to dc.

Now, let us consider the capacitor in the diagram. The formula for capacitive reactance shows that as the frequency increases, the capacitive reactance, X_C, decreases. Therefore, there will be less opposition to the flow of alternating current in the circuit by the capacitor as the frequency rises. Thus by the proper choice of inductance and capacitance values, we can obtain desired filtering actions. That is, we can design filters to either suppress or permit the passage of a certain frequency or band of frequencies. The detailed method of doing this will be discussed in succeeding pages.

Filter Principles

Electrical filters used in carrier systems, as stated before, consist of inductive and capacitive elements. These two elements must be employed together to form a filter that can differentiate between alternating currents of various frequencies. To clarify this point, let us consider the simplified circuit below.

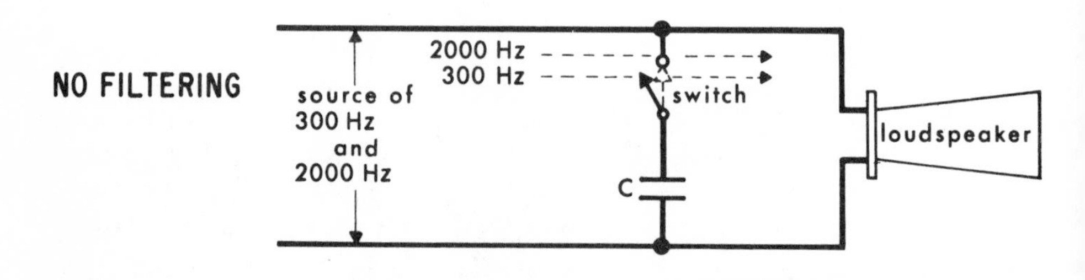

The loudspeaker or load in this circuit is shunted by a small capacitor. A constant-voltage source supplies frequencies of 300 and 2,000 Hz to the loudspeaker and capacitor load. If the switch to place the capacitor across the loudspeaker were open and closed, we would observe no effect on either the 300- or 2,000-Hz currents. In other words, the capacitor by itself performs no appreciable filtering action. Now, let us place an inductor, L, in series with the capacitor and loudspeaker load.

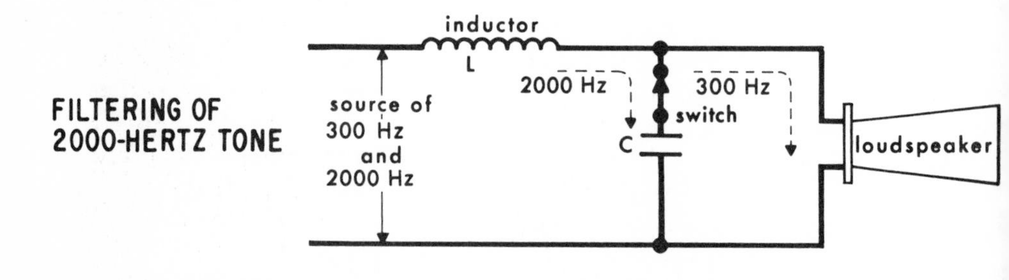

From the formula for inductive reactance, we note that the inductive reactance, X_L, is much greater at 2,000 Hz than at 300 Hz. The inverse is true for the capacitive reactance, X_C. This increased inductive reactance at 2,000 Hz will result in a relatively smaller amount of 2,000 Hz current flowing in the loudspeaker. We will hear the 300-Hz tone, but very little of the 2,000-Hz tone. This shows that the coil and capacitor form a filter having a much higher attenuation to the 2,000-Hz frequency than to 300-Hz.

Types of Filters

There are four types of electrical filters in general use in carrier systems. They are designated: low-pass; high-pass; bandpass; and band-rejection or band-stop.

Low-pass filters readily permit the passage of frequencies from about zero up to a predetermined cutoff frequency. All frequencies above this cutoff frequency are attenuated. Low-pass filters are used extensively, for example, to pass the 200-3,400-Hz voice frequency band of the demodulator output.

High-pass filters in this case, greatly attenuate frequencies in the band below about 3,400 Hz, but permit all higher frequencies to pass through. Examples of the use of low-pass and high-pass filters in carrier terminals were given in the discussion of *FDM Carrier System Fundamentals.*

Bandpass filters, as the name implies, permit the passage of only a certain band of frequencies between two specific frequencies. The frequencies below and above this pass band are greatly attenuated.

The diagram illustrates the effect of connecting low-pass and high-pass filters in series to form a bandpass filter.

Band-stop filters may be considered as the reverse of bandpass filters. They reject a band of frequencies located between two specific frequencies. The frequencies above and below this rejected stop band are permitted to pass through without attenuation.

Bandpass Filter Arrangement

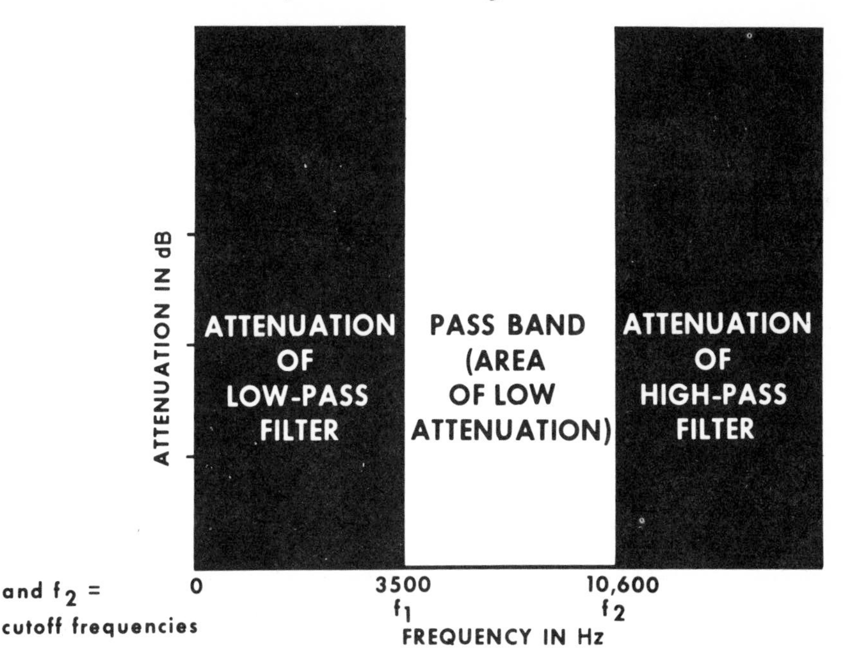

Series Resonant Filter Circuits

Before proceeding with the designs of the various filter types, it is important that we also understand the functions of resonant circuits. The inductor, L, and the capacitor, C, comprising a filter may be connected in series or in parallel. A series circuit in which the inductive and capacitive reactances are equal ($X_L = X_C$) is said to be a *resonant* or tuned circuit. It can be shown mathematically that a series resonant circuit theoretically has no impedance. This can only occur at a particular frequency as given by the formula below.

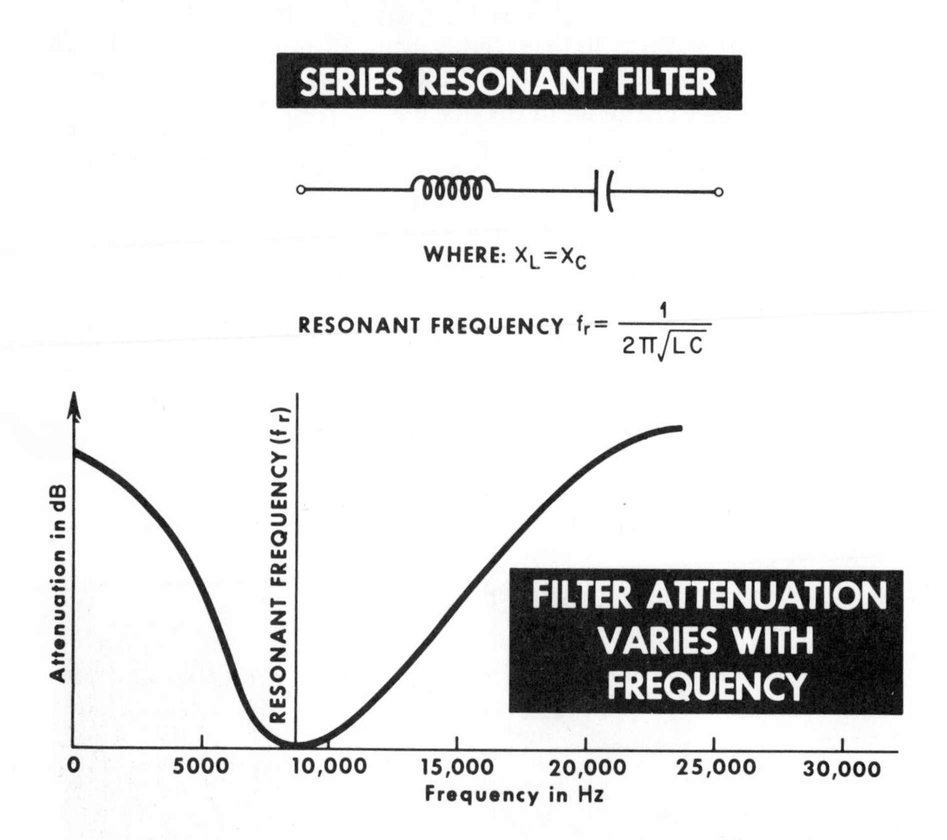

For example, assume that a filter consists of a coil having an inductance of 0.1 henry in series with a capacitance of 0.004 microfarad. By substituting in the above formula, we find that the resonant frequency, f_r, is 8,000 Hz.

This characteristic of series resonant circuits is an important factor in filter design. Series resonance permits the ready passage of ac (voice frequency or carrier currents) at the resonant frequency. This condition is illustrated in the graph which shows how the attenuation of a filter is affected by the resonant frequency.

Parallel Resonant Filter Circuits

When the inductive and capacitive elements in a filter are connected in parallel, a different circuit condition is obtained at the resonant frequency. The same formulas used for determining the series circuit resonance are applicable to the parallel circuit arrangement. The impedance at the resonant frequency, however, is very high (theoretically infinite). In the series resonant condition, it will be recalled, the impedance was zero.

The graph illustrates the parallel resonant circuit condition with respect to the attenuation at the resonant frequency. Observe the very high attenuation that occurs at the resonant frequency. We can say, therefore, that parallel resonance rejects or stops the passage of ac (voice frequency or carrier currents) at the resonant frequency. This condition is also used to a considerable extent in the design of filters for various applications in carrier systems.

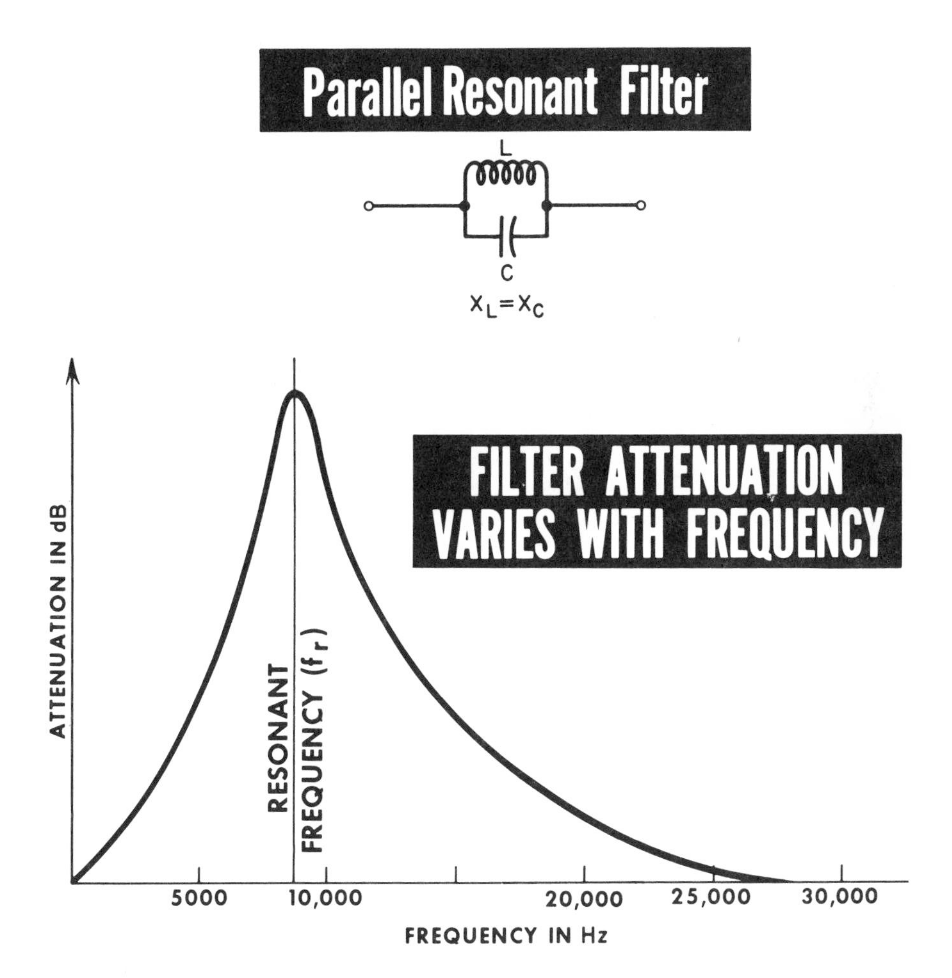

Filter Design Parameters

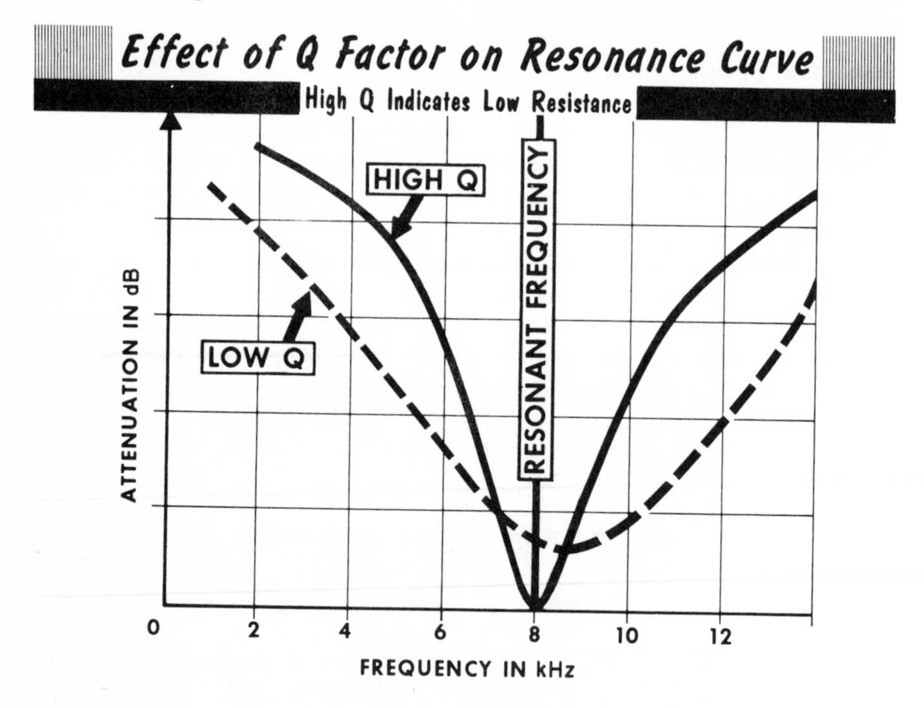

The shape of the resonance curves, as in the preceding graphs of series and parallel filters, determines to a considerable extent the filter's effectiveness. The sharpness of this curve as it approaches the resonant frequency depends on the resistance of the inductive and capacitive components of the filter. We can also say that the degree of resonance sharpness is determined by the relationship between the reactance and resistance of the filter elements. This relationship is designated *Q* or the *Quality Factor.* It is expressed by the formulas:

$$Q = \frac{X_L}{R} \text{ for a coil or inductor}$$

$$Q = \frac{X_C}{R} \text{ for a capacitor}$$

A large Q factor indicates an effective low resistance. This results in a steep resonance curve. A small Q factor indicates a high-resistance relationship in the filter's elements. Therefore, the resonance curve is less sharp and more rounded at the resonant frequency. The curves in the graph illustrate these two Q conditions. The use of high Q components is very desirable for filter elements. Typical desirable Q factors are from 200 to 300 for filters covering carrier frequencies up to about 150 kHz. It may be of interest that values of Q of 10,000 or greater can be obtained with crystal filters and active type electronic filters.

Design Parameters for Low-Pass Filter

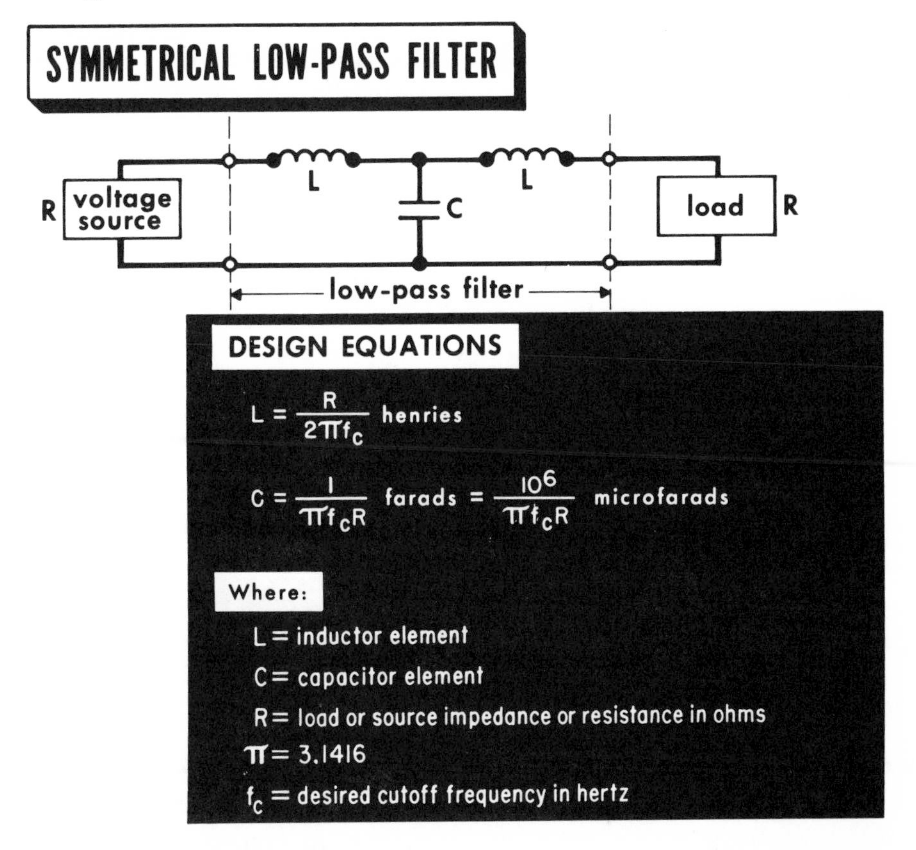

The precise values of the inductive and capacitive elements of an electrical filter, to provide satisfactory filtering action in a desired frequency range, can be computed mathematically. The means for accomplishing the necessary computations and some examples of filter design are shown on this and the following pages. The filters to be described are of very simple types, and they are presented only to illustrate the basic principles involved. The design of filter networks required for a given application is far more complicated and beyond the scope of this book. A great deal of careful mathematical calculation is required, and computers are often used for this purpose.

In designing a low-pass filter, there should be a symmetrical arrangement of the filter elements. This is done to present the same impedance value at both terminations of the filter. The above schematic illustrates a simple symmetrical low-pass filter. This symmetrical network is obtained by connecting two required coils (inductors) in series and connecting the required capacitor to the centertap of the coils. The design equations are also given in the diagram. Note that the cutoff frequency and the value of the load or source resistance (impedance) must be known.

Example of Low-Pass Filter Design

Let us now follow closely the necessary computations for calculating the inductive and capacitive elements of a simple low-pass filter. It is assumed that this filter will be used in the output circuit of a demodulator. For design purposes, the carrier circuit or load impedance is assumed to be 600 ohms and the cutoff frequency to be 3,400 Hz. By substituting the above values in the previously given design equations for a low-pass filter, we obtain:

$$L = \frac{R}{2\pi f_c} = \frac{600}{6.28(3{,}400\text{ Hz})} = 0.028 \text{ henry}$$

$$C = \frac{1}{\pi f_c R} = \frac{10^6}{3.14(3{,}400 \times 600)} = 0.156 \text{ microfarad}$$

Referring to the schematic diagram, the above calculations indicate that each series inductor L, has a value of 28 millihenries. Similarly, the shunt capacitor, C, has a calculated value of 0.156 microfarad. Note that the inductive element is in series with the load and that the capacitive element is in shunt with the circuit load. This is one of the fundamental requirements for a low-pass filter. The resultant frequency characteristics of a typical low-pass filter are illustrated in the graph.

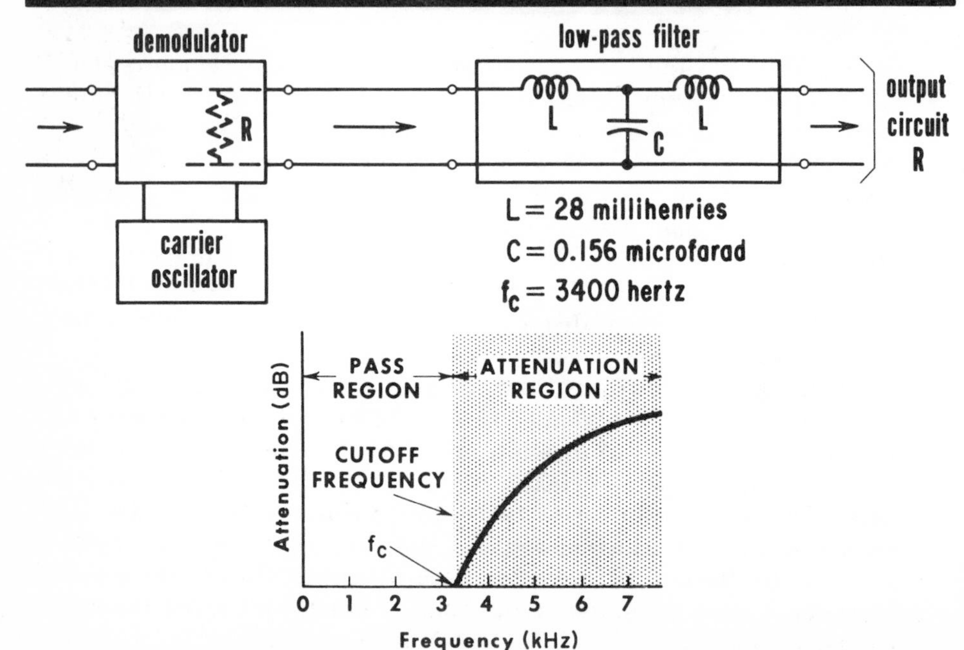

Design Parameters of the High-Pass Filter

The connections of the elements comprising a high-pass filter differ from those used in a low-pass filter. The capacitive element in the high-pass filter is placed in series with the load; the inductive element is connected in parallel with the circuit load. The impedance seen or presented, therefore, is the result of the capacitance in series with the parallel combination of the load and the inductive element. The simplified schematic below also gives the design equations.

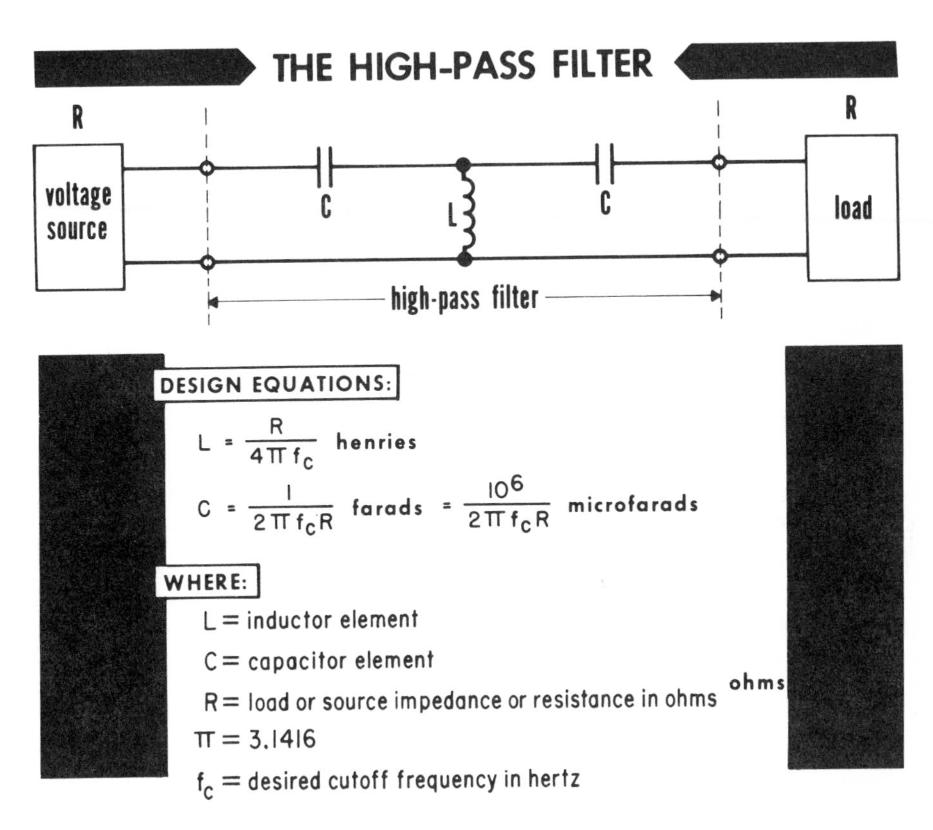

It will be recalled that the voltage across the elements of a series circuit is directly proportional to the value of the resultant impedance. Also, as previously discussed, the capacitive reactance, X_C, is much greater at a low frequency, such as 250 Hz, than at the higher frequencies, say 2,000 Hz. The inverse order is true for the inductive reactance, X_L. Therefore, the ratio of the capacitive reactance, X_C, to the total impedance, Z, of the parallel combination will be much higher at the low 250-Hz frequency than at the higher 2,000 Hz. This causes a relatively small amount of low-frequency 250-Hz current to flow in the load. It is desirable that high-pass filters also be symmetrical. This is usually accomplished by placing an equivalent series capacitance between the shunt inductance and the load.

Example of High-Pass Filter Design

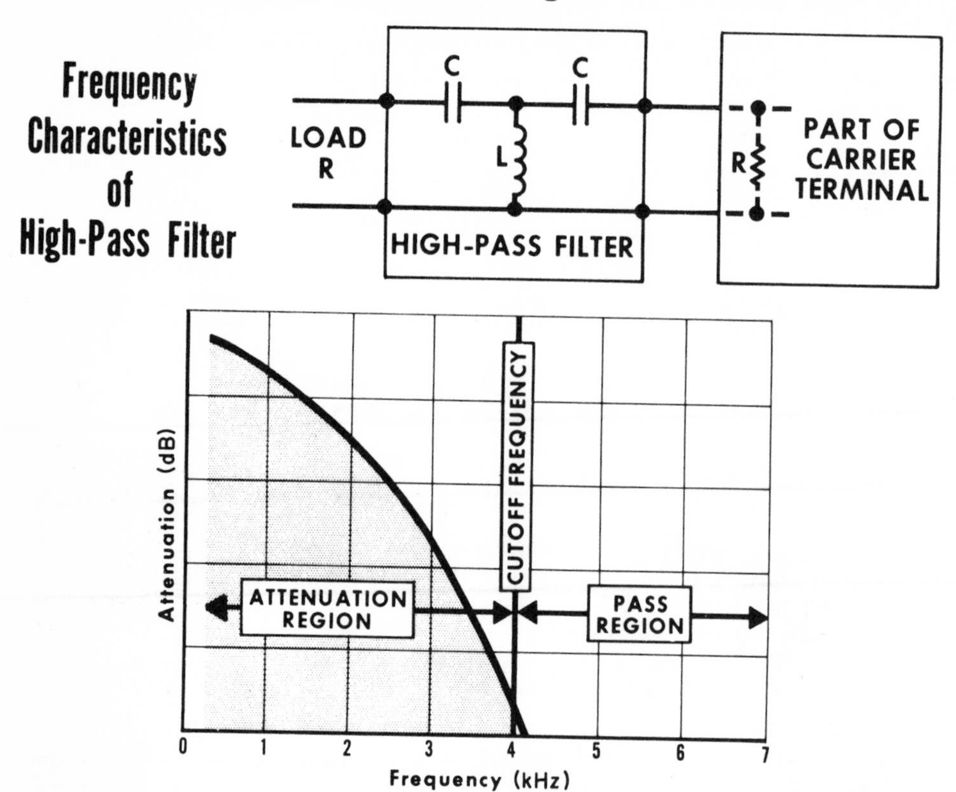

The mathematical calculations for designing a high-pass filter follow a pattern similar to that of the low-pass filter. In the following example, it is assumed that a high-pass filter is needed to attenuate the voice-frequency currents that may be present in a part of a carrier terminal. The load impedance, R, is taken as 600 ohms. The cutoff frequency, fc, is to be 4,000 Hz. These values are now substituted in the previously given equations for the design of a high-pass filter. The following results are obtained:

$$L = \frac{R}{4\pi f_c} = \frac{600}{12.56(4{,}000)} = 11.9 \text{ millihenries}$$

$$C = \frac{10^6}{2\pi f_c R} = \frac{10^6}{6.28(4{,}000 \times 600)} = 0.066 \text{ microfarad for each capacitor}$$

The capacitive element is in series with the load in the high-pass filter. The inductive element is in parallel with the load. These are the fundamental requirements in the design of a high-pass filter. The frequency characteristics of the above designed high-pass filter are also shown in the graph.

Design Parameters of Bandpass Filter

A bandpass filter may be formed, as previously shown by connecting a low-pass filter in series or cascade with a high-pass filter. There are two attenuation regions in a bandpass filter separated by the desired pass band. Each attenuation region has its particular cutoff frequency. Therefore, one attenuation region extends from about zero frequency to the low cutoff frequency, f_1. The other attenuation region ranges from the high cutoff frequency, f_2, to, theoretically, infinite frequency. The pass band lies between cutoff frequencies, f_1 and f_2, as shown.

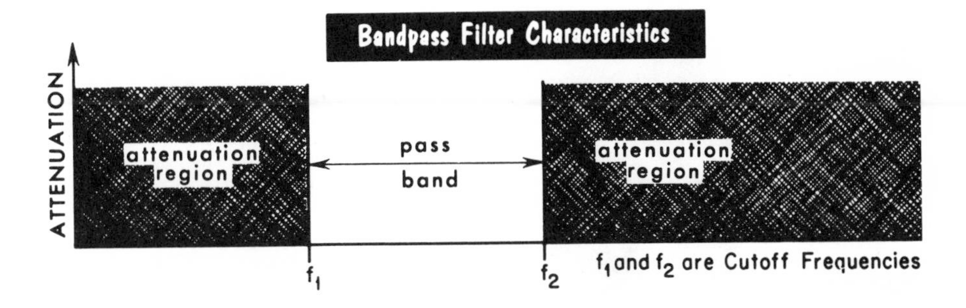

The prime design parameters of a bandpass filter are a low-pass and a high-pass filter which have overlapping cutoff frequencies. These filters can be arranged in cascade connection as in schematic diagram A. This is a workable bandpass filter. A more commonly used circuit arrangement is shown in diagram B.

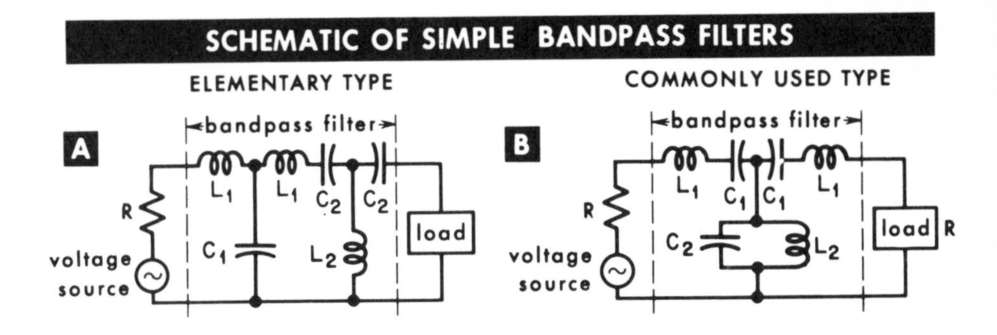

In diagram B, each series branch of the filter consists of an inductor and capacitor in series. The shunt branch of the filter has an inductor and capacitor in parallel connection. The entire filter is a symmetrical network. It is inserted between matched impedances, as for example, in the 600-ohm transmitting branch of a carrier terminal and the 600-ohm telephone transmission line or cable pair.

Design Equations for Bandpass Filter

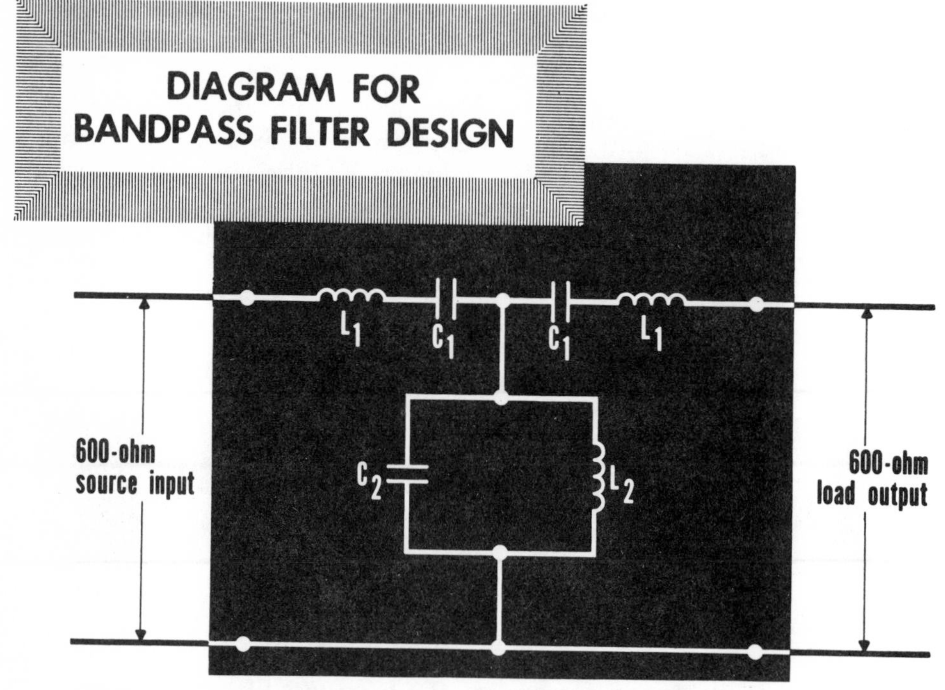

The design of a bandpass filter follows, in general, the equations previously given for the low-pass and high-pass filters. It is necessary to know the value of each desired cutoff frequency, f_1 and f_2, and the value of the load resistance or impedance to be matched.

$$L_1 = \frac{R}{2\pi(f_2 - f_1)} \text{ henry}$$

$$L_2 = \frac{(f_2 - f_1)\,R}{4\pi f_1 f_2} \text{ henry}$$

$$C_1 = \frac{(f_2 - f_1)\,10^6}{2\pi f_1 f_2\,R} \text{ microfarad}$$

$$C_2 = \frac{10^6}{\pi(f_2 - f_1)R} \text{ microfarad}$$

Where:

L_1 = Series inductive elements
L_2 = Shunt inductive elements
C_1 = Series capacitive elements
C_2 = Shunt capacitive elements

f_1 = Low cutoff frequency
f_2 = High cutoff frequency
R = Load impedance or source resistance

Example of Bandpass Filter Design

In the following example, the design of a bandpass filter is calculated for the output circuit of a modulator in a carrier terminal. The known data are: load impedance is a 600-ohm circuit; desired pass band is between 4,000 and 12,000 Hz. Therefore, $f_1 = 4{,}000$ Hz and $f_2 = 12{,}000$ Hz. We will now substitute the above given values in the previously stated design equations for a bandpass filter:

$$L_1 = \frac{R}{2\pi(f_2 - f_1)} = \frac{600}{6.28(12{,}000 - 4{,}000)} = 11.9 \text{ millihenries}$$

$$L_2 = \frac{(f_2 - f_1)\,R}{4\pi f_1 f_2} = \frac{(12{,}000 - 4{,}000)600}{12.56(4{,}000 \times 12{,}000)} = 7.9 \text{ millihenries}$$

$$C_1 = \frac{(f_2 - f_1)\,10^6}{2\pi f_1 f_2\,R} = \frac{(12{,}000 - 4{,}000)\,10^6}{6.28(4{,}000 \times 12{,}000)600} = 0.044 \text{ microfarad}$$

$$C_2 = \frac{10^6}{\pi(f_2 - f_1)\,R} = \frac{10^6}{3.14(12{,}000 - 4{,}000)600} = 0.066 \text{ microfarad}$$

The resultant values are:

Series inductor coils, L_1 = 11.9 millihenries
Series capacitors, C_1 = 0.044 microfarad
Shunt inductor coil, L_2 = 7.9 millihenries
Shunt capacitor, C_2 = 0.066 microfarad

The above calculated values will provide a pass band of from 4,000 to 12,000 Hz. The frequency characteristics of this designed bandpass filter are illustrated below:

FREQUENCY CHARACTERISTICS OF BANDPASS FILTER

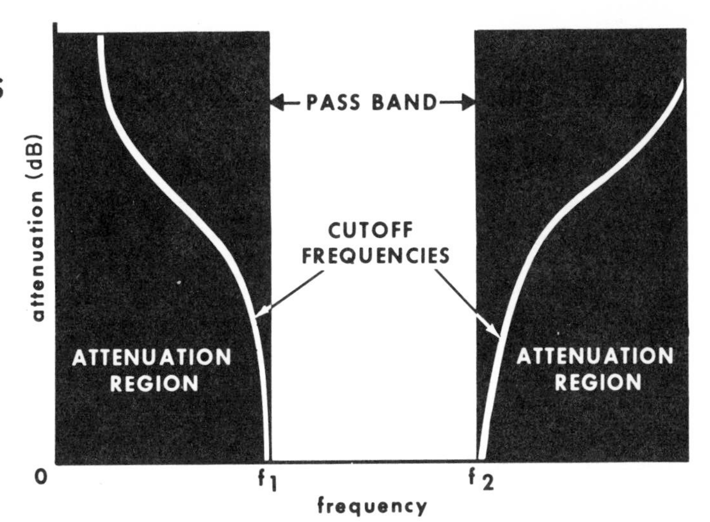

CUTOFF FREQUENCIES

f_1 = 4000 Hertz
f_2 = 12,000 Hertz

Design Parameters of Band-Rejection Filter

The band-rejection or band-stop filter has one attenuation region between two pass bands. The stop-band or attenuation region is bounded by two cutoff frequencies. They are usually designated f_1 and f_2 for the low-frequency and high-frequency borders, respectively. The graph illustrates the frequency characteristics of a simple band-rejection filter.

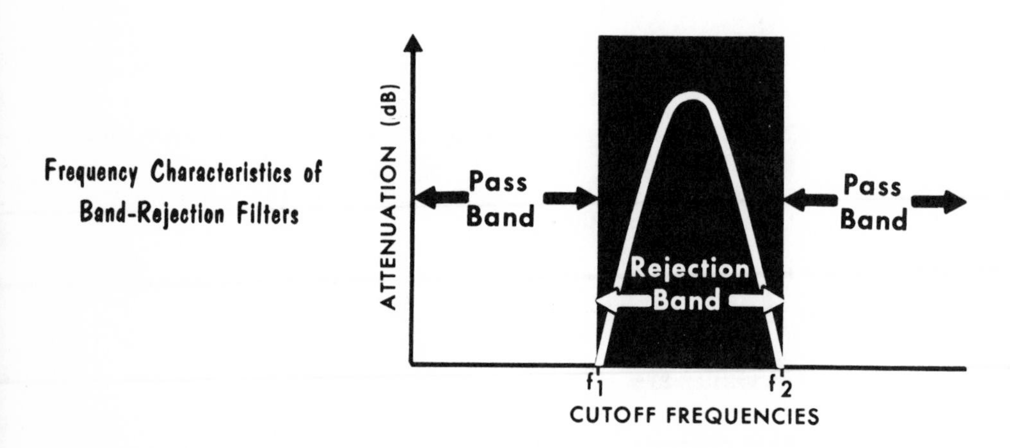

Band-rejection filters usually employ a Pi-type connection of the filter elements. The schematic diagram below shows a common circuit arrangement of a band-rejection filter:

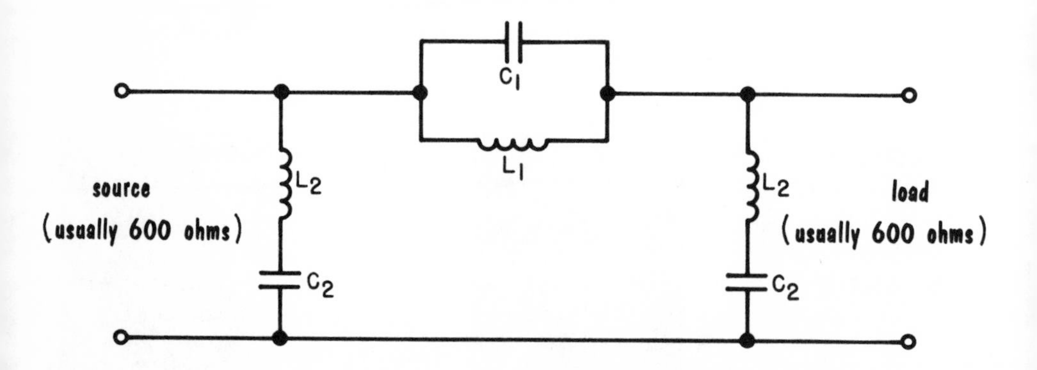

SCHEMATIC OF BAND-REJECTION FILTER PI-TYPE

Note that there are three separately tuned circuits. The parallel tuned circuits is inserted between the two series tuned circuits which, in turn, are shunted across the load.

Design Equation and a Band-Rejection Filter Design

The following are typical design equations of a band-rejection filter of the Pi-type:

$$L_1 = \frac{(f_2 - f_1)\,R}{\pi f_1 f_2} \text{ henries} \qquad C_1 = \frac{10^6}{4\pi(f_2 - f_1)\,R} \text{ microfarads}$$

$$L_2 = \frac{R}{2\pi(f_2 - f_1)} \text{ henries} \qquad C_2 = \frac{(f_2 - f_1)\,10^6}{2\pi f_1 f_2\,R} \text{ microfarads}$$

L_1 and C_1 comprise the series arm and L_2 and C_2 are the shunt arm elements as in the case of the bandpass filter.

Now, let us design a band-stop filter for a 600-ohm circuit with a stop or rejection band from 3,000 to 9,000 Hz. By substituting these given values in the above equations, we can obtain the following design values:

$$L_1 = \frac{(9{,}000 \times 3{,}000)600}{3.14(3{,}000 \times 9{,}000)} = 42.4 \text{ millihenries}$$

$$L_2 = \frac{600}{6.28(9{,}000 - 3{,}000)} = 15.9 \text{ millihenries}$$

$$C_1 = \frac{10^6}{12.56(9{,}000 - 3{,}000)600} = 0.022 \text{ microfarad}$$

$$C_2 = \frac{(9{,}000 - 3{,}000)\,10^6}{6.28\,(3{,}000 \times 9{,}000)600} = 0.059 \text{ microfarad}$$

Questions and Problems

1. What functions are performed by electrical filters?
2. What electrical components comprise a filter?
3. Give the formulas for inductive and capacitive reactances.
4. What general types of filters are employed in carrier systems?
5. What is the theoretical impedance of a series resonance circuit? Why?
6. A series resonant filter includes an inductance of 20 millihenries and a capacitance of 0.045 microfarads. What is its resonant frequency?
7. What is the effect of a parallel resonant filter circuit?
8. Draw the basic schematic diagrams of low-pass and high-pass filters.
9. A low-pass filter has a cutoff frequency of 3,400 Hz and a load impedance of 600 ohms. What are its inductance and capacitance values?
10. What are the basic elements of a bandpass filters?

Carrier Terminals

The various components of each carrier channel may be mounted on subassemblies, plug-in units, or a separate chassis. The complete assembly, consisting of two or more carrier channels and associated common equipment. power supplies, pilot, and alarm circuits, etc., is called a *carrier terminal.* This terminal provides the necessary channeling and multiplexing equipment for the required number of voice circuits to be served by the particular carrier system. The two principal types, classified by the transmission facility, are *two-wire* and *four-wire* carrier terminals. Two-wire carrier terminals, as the name implies, utilize a single cable pair for both directions of transmission. Four-wire terminals connect to two separate cable pairs, one for each direction of transmission. Carrier terminals for cable and radio systems invariably are of the four-wire type. Two-wire carrier terminals are normally connected to open-wire lines and cable circuits for extending telephone service to subscribers in rural areas. The GTE Lenkurt #82A station carrier system is an example.

Operation of Four-Wire Carrier Terminal

In order to understand carrier operations, let us first analyze an early open-wire carrier system known as type C. It had a capacity of one voice circuit and three superimposed carrier channels. A simplified diagram of this single sideband suppressed carrier system, as used with a toll trunk in past years, appears on page 104. Note that the two-wire side of each channel connects to the toll trunk. The four-wire side goes through the common amplifier, the equalizer, and the bandpass filters. Its output connects to a four-wire telephone line in which one pair is used for each direction of transmission. The diagram shows the VF channel and one carrier channel of the type C terminal. Additional carrier channels up to three may be paralleled as shown in the diagram. Each channel has its own *transmitting* and *receiving branch*; a signal will now be traced through it to show how it operates.

Carrier Transmitting Branch. From toll trunk No. 2, the speech or voice signals are routed to the hybrid network of the carrier channel. A voice frequency (20/1,000 Hz) ringer or other equipment such as to provide a suitable frequency for signaling purposes was also furnished, but they are not considered in this discussion. The speech signals are directed to the carrier transmitting branch by the hybrid arrangement, and enter the pad circuit. The signal voltage is reduced to a predetermined value by the pad. This is done so that the signal voltage reaching the modulator will be less than that from the carrier frequency oscillator. From the pad, the voice signals enter the transmitting low-pass filter which is designed to reject all frequency components above about 3,400 Hz. This will limit the voice bandwidth to 200-3,400 Hz for more efficient transmission. The signals now go to

Operation of Four-Wire Carrier Terminal (cont.)

the modulator where they are combined with the 12,900-Hz output of the carrier frequency oscillator. As a result of the modulation process, the output of the modulator will contain, along with other frequencies, the upper (13,100-16,300 Hz) and lower (9,500-12,700 Hz) sidebands. The 12,900-cycle carrier frequency is suppressed within the modulator. These various frequency components now enter the bandpass filter which permits only the upper sideband to go through. The other frequencies are attenuated. The upper-sideband carrier signals now pass into the input circuit of the *common transmitting amplifier.*

At this point, note that the voice frequency channel's *transmitting branch* also is connected to the input of the common transmitting amplifier. However, the low-pass filter in the voice frequency channel's transmitting branch prevents the 13,100-16,300-Hz signals of the carrier channel from entering the voice frequency circuit. Therefore, both the voice frequency and carrier channel signals can enter only the common transmitting amplifier. The signals are raised by this amplifier to the proper level for transmission over the telephone line to the distant East carrier terminal.

Carrier Receiving Branch. The carrier signals entering the West carrier terminal from the distant East end will be at low level because of the line attenuation. Assume that they are in the 9,500-12,700-Hz frequency range. This is the lower sideband of the East carrier terminal's modulator output. These carrier signals, after passing through the equalizer network, are amplified by the *common receiving amplifier.* They will next see two parallel paths—the low-pass filter in the voice frequency receiving branch, and the bandpass filter in the carrier channel receiving branch. Since the carrier signals are in the 9,500-12,700-Hz range, they will be rejected by the low-pass filter because it can pass only 200-3,400-Hz frequencies. Therefore, the carrier signals will be directed to the bandpass filter. The pad circuit after the bandpass filter is set to reduce the carrier signal voltage so that it will enter the demodulator at a lower value than that from the carrier frequency oscillator. The demodulator output contains the original voice intelligence along with other combinations of the carrier oscillator frequency and the incoming carrier signal. The receiving low-pass filter following the demodulator allows only reproduced original speech signals (in the 200-3,400-Hz range) to pass to the voice frequency amplifier. This amplifier increases the voice signal level for further transmission through the hybrid network to the toll trunk circuit.

The operations of the second and succeeding carrier channels follow a similar pattern. Different carrier oscillator frequencies are generated for each carrier channel. This was discussed previously under Carrier Frequency Allocations. For similar reasons, the respective bandpass filters of each carrier channel are designed to permit the passage only of the frequency range assigned to that carrier channel.

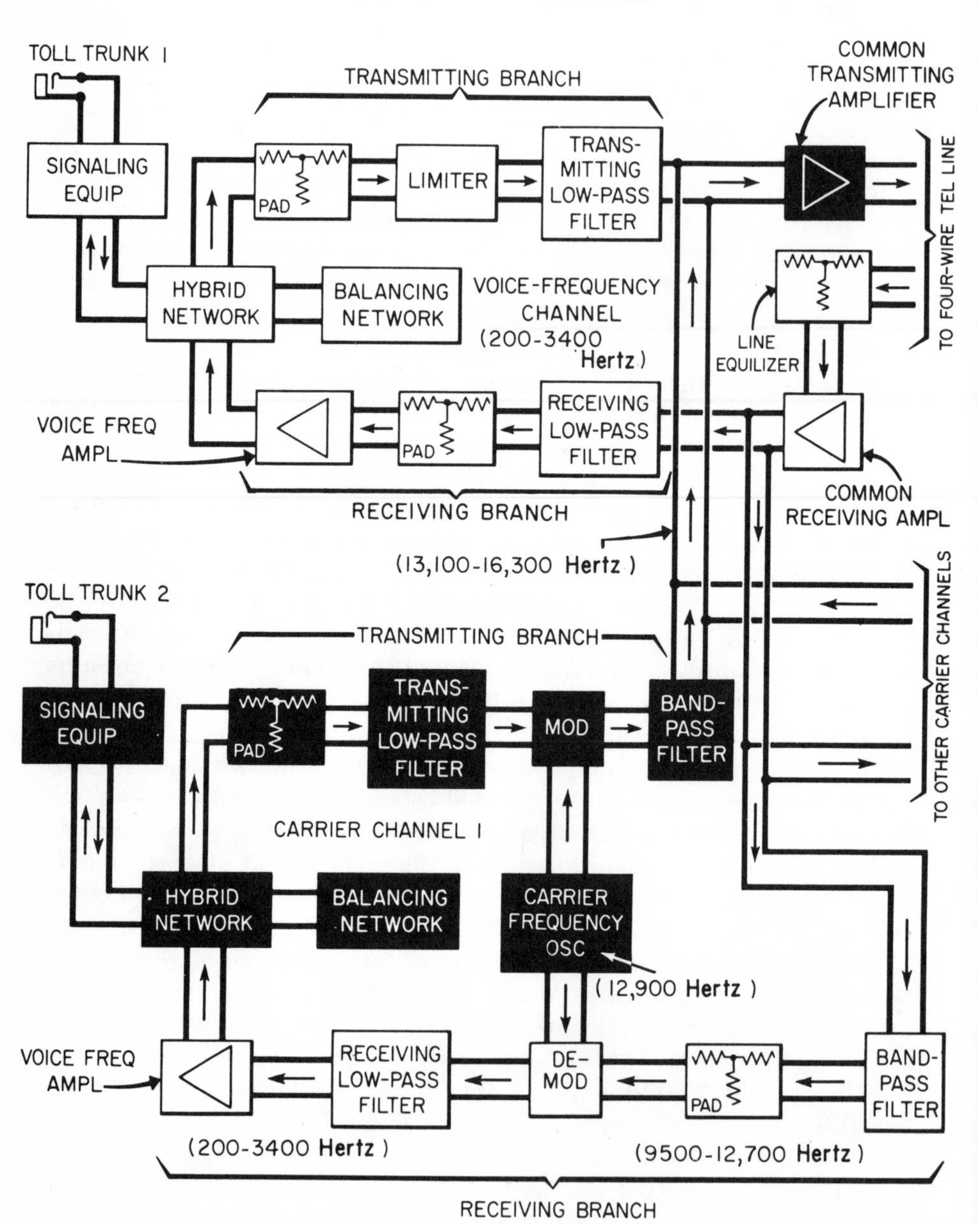

TYPICAL FOUR-WIRE CARRIER TERMINAL

Operation of Voice Channel in Four-Wire Carrier Terminal

Although the voice channel is not used in large open-wire carriers nor in cable carrier systems, knowing its functions may better your grasp of multiplex techniques. In the diagram (opposite), follow the signals from toll trunk No. 1 through the associated voice channel.

Voice Frequency Transmitting Branch. The transmitting branch of the voice channel of this carrier terminal differs from the carrier channels. It does not contain a modulator, carrier frequency oscillator, or bandpass filters—essential elements of a carrier channel. The voice channel comprises, in addition to the usual hybrid arrangement, a pad, transmitting limiter and low-pass filter, as illustrated in the block diagram. The voice signals from the trunk circuit are routed by the hybrid network to the pad in the transmitting branch. The pad circuit reduces the voice signal level so that signals from all channels reach the common transmitting amplifier at about the same strength. This also prevents the possibility of overloading the amplifier. The limiter circuit absorbs momentary large peak voltages from speech signals which could overload the common transmitting amplifier.

The transmitting low-pass filter, which follows the limiter, restricts the bandwidth to below 3,400 Hz as in the case of the carrier channel. The voice frequency signals now enter the common transmitting amplifier in parallel with the output from the carrier channels. The voice frequency signals, however, cannot go through the carrier transmitting branch because of its bandpass filter. The common transmitting amplifier raises the level of the voice frequency channel signal to the desired value for transmission to the distant East carrier terminal.

Voice Frequency Receiving Branch. The receiving branch of the voice frequency channel also is different from that of the carrier channel. It does not have a demodulator, carrier frequency oscillator or bandpass filter. It does contain a receiving low-pass filter, pad, and voice frequency amplifier. The incoming voice frequency channel signals are at low level in the 200-3,400-Hz range. They are amplified by the common receiving amplifier to the same levels as the carrier channel signals.

There are now two possible paths for the voice frequency signals. The path to the carrier receiving branch, however, is blocked by the bandpass filter. Therefore, the voice signals can pass unhindered only through the receiving low-pass filter in the voice frequency receiving branch. The pad following this low-pass filter reduces the signal to a value that will not overload the connecting voice frequency amplifier. This amplifier increases the level of the voice signals for subsequent transmission through the hybrid network and the connecting trunk circuit to the toll center. The diagram on page 104 used as a basis of our analysis represents a typical four-wire carrier terminal. There are any number of variations but functionally they are very much alike.

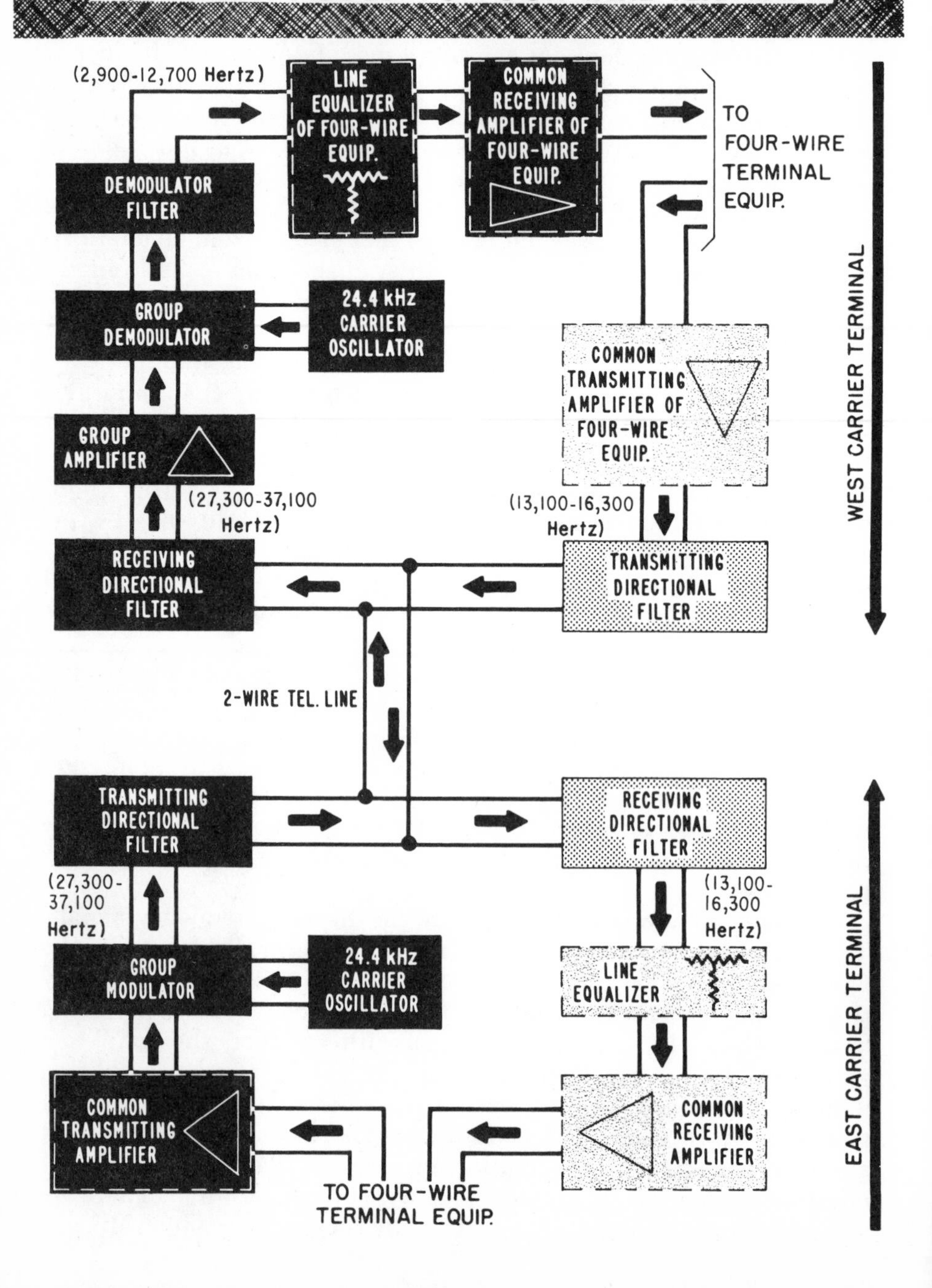
TWO-WIRE CARRIER SYSTEM AS RESULT OF EQUIPMENT ADDED TO FOUR-WIRE TERMINAL
(2,900-12,700 Hertz)
LINE EQUALIZER OF FOUR-WIRE EQUIP.
COMMON RECEIVING AMPLIFIER OF FOUR-WIRE EQUIP.
TO FOUR-WIRE TERMINAL EQUIP.
DEMODULATOR FILTER
GROUP DEMODULATOR
24.4 kHz CARRIER OSCILLATOR
COMMON TRANSMITTING AMPLIFIER OF FOUR-WIRE EQUIP.
GROUP AMPLIFIER
(27,300-37,100 Hertz)
(13,100-16,300 Hertz)
RECEIVING DIRECTIONAL FILTER
TRANSMITTING DIRECTIONAL FILTER
WEST CARRIER TERMINAL
2-WIRE TEL. LINE
TRANSMITTING DIRECTIONAL FILTER
RECEIVING DIRECTIONAL FILTER
(27,300-37,100 Hertz)
(13,100-16,300 Hertz)
GROUP MODULATOR
24.4 kHz CARRIER OSCILLATOR
LINE EQUALIZER
COMMON TRANSMITTING AMPLIFIER
COMMON RECEIVING AMPLIFIER
TO FOUR-WIRE TERMINAL EQUIP.
EAST CARRIER TERMINAL

Two-Wire Carrier Terminal Operations

The carrier channel operations just described were for a typical four-wire carrier terminal that connects to two open-wire pairs, or to two pairs in a toll cable. Separate pairs are used for each direction of transmission. When the same open-wire pair is to be used for both directions of transmission, it is necessary to provide additional equipment in the four-wire carrier terminal. This equipment is needed to convert the terminal for two-wire operations. The additional apparatus is required primarily to change the carrier frequency range for one direction of transmission, because the same carrier frequencies cannot be transmitted over the same wire pair from the West and East terminals; the carrier signals would interfere with each other.

The major additional items needed are: directional filters; group modulator; group demodulator; group filter; and group amplifier. The block diagram on the opposite page shows how these items are used to convert a four-wire carrier terminal to two-wire operations. The transmitting and receiving directional filters in this diagram are bandpass filters which separate the transmitting and receiving branches at each carrier terminal. It is possible to use hybrid networks for this purpose, but they are not as efficient or suitable for operation over a wide frequency range. Note that the equipment is not the same at each carrier terminal. The West transmitting directional filter and the East receiving directional filter are identical. Similarly, the East transmitting and West receiving directional filters are equivalent, to pass the same band of frequencies. A detailed description of operations follows (refer to diagram on page 106).

West-East Carrier Transmissions. The output band of frequencies (13,100-16,300 Hz) from the common transmitting amplifier in the West carrier terminal goes through the transmitting directional filter. As previously stated, the two directional filters at a carrier terminal do not pass the same band of frequencies. Therefore, these carrier signals are rejected by the receiving directional filter. This causes them to enter the two-wire transmission line and pass along to the distant East carrier terminal. Upon arrival at this terminal, the carrier signals find two possible paths: one through the receiving directional filter, and the other through the transmitting directional filter. The two directional filters at a carrier terminal do not pass the same frequency bands. The transmitting directional filter at the West carrier terminal and the receiving directional filter at the East terminal are identical, as noted above; so the carrier signals can pass only through the receiving directional filter to the common receiving amplifier of the East carrier terminal. The succeeding operations are the same as those described for the four-wire terminal operations.

East-West Carrier Transmissions. It is necessary, for the reasons stated, that the E-W carrier frequency band be different from that of the W-E carrier transmissions on a common two-wire line. A group

Two-Wire Carrier Terminal Operations (cont.)

modulator and a higher carrier frequency oscillator provide this frequency change at the East carrier terminal. The group modulator is essentially the same device as the modulator used in each carrier channel except that it operates at a higher carrier frequency. The output of the common transmitting amplifier, which contains the several carrier channel frequencies, connects to the group modulator. As the result of the modulation process, all the channel frequencies are superimposed on the new high carrier frequency in the group modulator, which must be well above any of the input carrier channel frequencies. For example, assume that the East carrier terminal is equipped for three carrier channels with these output lower-sideband frequencies:

Channel No.	*Carrier Oscillator Frequency*	*Resultant Lower-Sideband Signals*
1	9,400 Hz	6,000-9,200 Hz
2	6,300 Hz	2,900-6,100 Hz
3	12,900 Hz	9,500-12,700 Hz

The above band of frequencies in the range of 2,900 to 12,700 Hz are amplified by the common transmitting amplifier in the East carrier terminal and then sent to the group modulator. If we assume further that the group carrier frequency oscillator is 24,400 Hz, the following upper-sideband frequencies will result from the modulation process, excluding all other components:

Channel No.	*Original Channel Frequencies*	*Resultant Group Modulated Frequencies*
1	6,000-9,200 Hz	30,400-33,600 Hz
2	2,900-6,100 Hz	27,300-30,500 Hz
3	9,500-12,700 Hz	33,900-37,100 Hz

The East terminal directional transmitting filter is designed to permit only the 27,300 to 37,100-Hz frequency band, as detailed above, to pass through to the two-wire transmission line. These carrier signals are rejected by the receiving directional filter of the East terminal because the two directional filters of a carrier terminal do not pass the same frequency band.

At the West carrier terminal, these E-W transmitted carrier signals (27,300-37,100 Hz) are rejected by the transmitting directional filter but accepted by the receiving directional filter. The signals now pass through a group receiving amplifier so that their level may be increased for the demodulation process. The group demodulator, which follows the group amplifier, has the same local generated carrier frequency of 24,400 Hz as was used in the group modulator in the East

Two-Wire Carrier Terminal Operations (cont.)

carrier terminal. Thus, the combining of the incoming carrier frequencies with the 24,400-Hz generated carrier in the group demodulator reproduces the same upper- and lower-sideband carrier channel frequencies that were originally generated at the East carrier terminal. (See table opposite.) Other frequency components are also produced but the demodulation filter after the group demodulator attenuates them. Only the original channel carrier frequencies can pass through to the common receiving amplifier in the West carrier terminal. The subsequent operations are the same as were described for the four-wire carrier terminal.

Four-Wire Carrier Frequency Repeaters

Carrier frequency transmissions over cable pairs are attenuated considerably more than voice frequency signals. This effect is due to the cable characteristics, as explained in the section on Cable Carrier Systems. Carrier terminals furnish a higher signal level to the wire pair than do voice frequency circuits. For example, carrier terminals of the 12-channel Western Electric N1 and the GTE Lenkurt 47A/N1 carrier systems transmit signal levels of +3 dBm (0.002 Watts) to +12 dBm (0.016 Watts) to the cable pair. This level is adequate for about 8 miles of standard No. 19 gage toll cable; after the 8 mile limit, a *carrier repeater* is necessary. Carrier repeaters are very similar to the voice frequency repeaters, described earlier. They are employed in both two-wire and four-wire carrier systems. The main elements of a four-wire carrier repeater are illustrated in the block diagram.

The amplifier, line equalizer, and associated equipment are identical for each direction of transmission. This is because separate wire pairs are used for four-wire circuits. Automatic regulation of the repeater gain by pilot carrier is usually provided in modern carrier systems. Carrier repeaters are normally inserted at equal distances in the telephone line between carrier terminals. The carrier repeater oper-

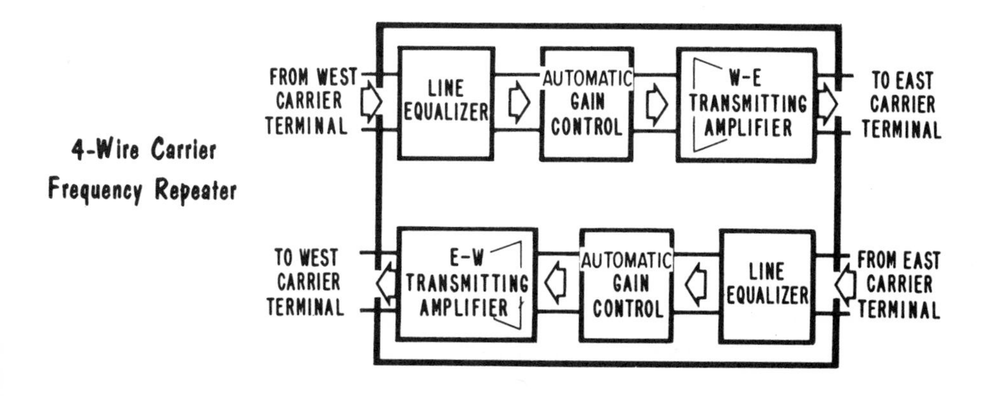

4-Wire Carrier Frequency Repeater

Four-Wire Carrier Frequency Repeaters (cont.)

ations are self explanatory. Briefly, the amplifier in each transmitting branch has 35-40-dB gain. The incoming carrier signals are raised to a preset level for further transmission to the next carrier repeater or the distant carrier terminal. The line equalizer makes up for unequal frequency attenuation in the cable pairs and adds attenuation to make the total attenuation or loss equal for all carrier frequencies received.

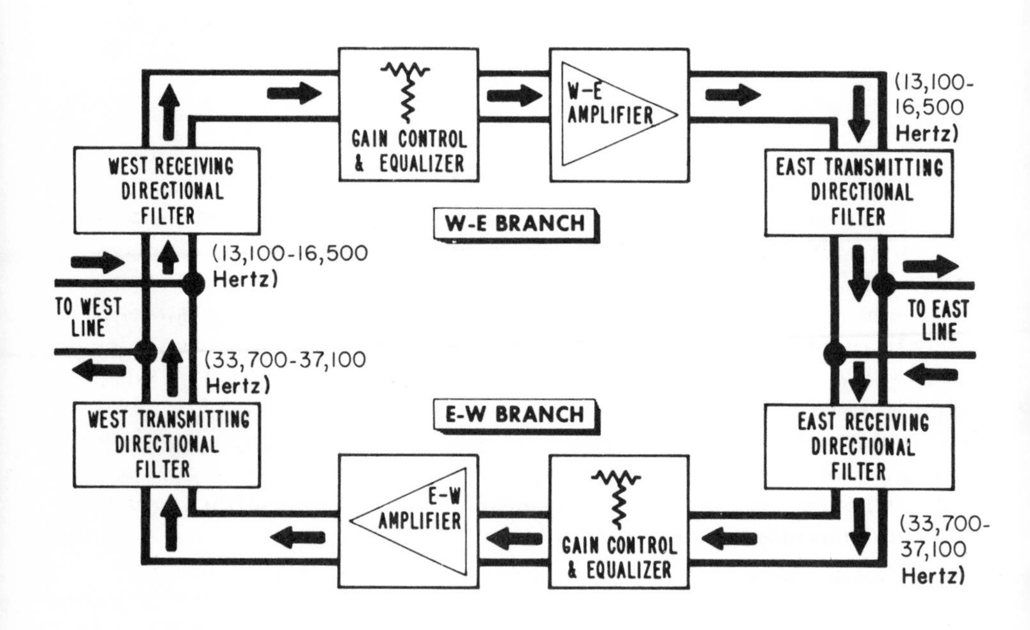

A two-wire carrier repeater is essentially the same as the four-wire type except that it includes four directional filters. These directional filters separate the sending and receiving paths. They are similar to those described for the two-wire carrier terminals. The West receiving and East transmitting directional or bandpass filters are identical, and the East receiving and West transmitting directional filters are the same. The block diagram shows a typical arrangement of a two-wire carrier frequency repeater with its essential components.

Two different carrier frequency bands are employed on a two-wire or single-pair open-wire line or cable pair. One frequency band is used for transmission in one direction and the second band for transmission in the other direction. For example, assume that a 13,100-16,300-Hz carrier signal enters the carrier repeater from the West carrier terminal. The West transmitting directional filter rejects this frequency band. The West receiving directional filter, however, accepts this signal and passes it along through the equalizer and gain control circuit to the W-E amplifier. There, the signal is amplified for further

Four-Wire Carrier Frequency Repeaters (cont.)

transmission through the East transmitting directional filter to the line. Similarly, carrier frequency signals from the East carrier terminal enter only the East receiving directional filter. These signals are amplified and sent through the West transmitting directional filter on to the line to the distant West carrier terminal or other carrier repeater. Carrier repeaters, as previously mentioned, often contain equipment for automatic gain control. Suppressed-carrier or pilot-carrier transmissions may be utilized for this purpose, as described later.

Carrier Frequency Control

In the foregoing discussions on carrier system operations, a particular carrier frequency was utilized for initiating the modulation process in each carrier channel. The same carrier frequency was employed at the demodulator in the receiving terminal for demodulation purposes. In the usual single-sideband suppressed-carrier system the channel carrier frequency is suppressed in the balanced modulator circuit. Therefore, no carrier frequency is transmitted to the distant receiving terminal to control the demodulation process. This means that the exact carrier frequency must be supplied at the receiving terminal to provide for proper demodulation of the channel carrier signals. Any difference in the carrier frequencies employed for modulation and demodulation will cause distortion and noise in the resultant demodulated speech signals.

For example, one channel of the West transmitting terminal has a modulation frequency of 12,900 Hz. The demodulator's carrier frequency at the East receiving terminal is assumed to be 12,880 Hz, a difference of 20 Hz. A speech signal in the 200-3,400 Hz band transmitted by this carrier channel would be demodulated at the East re-

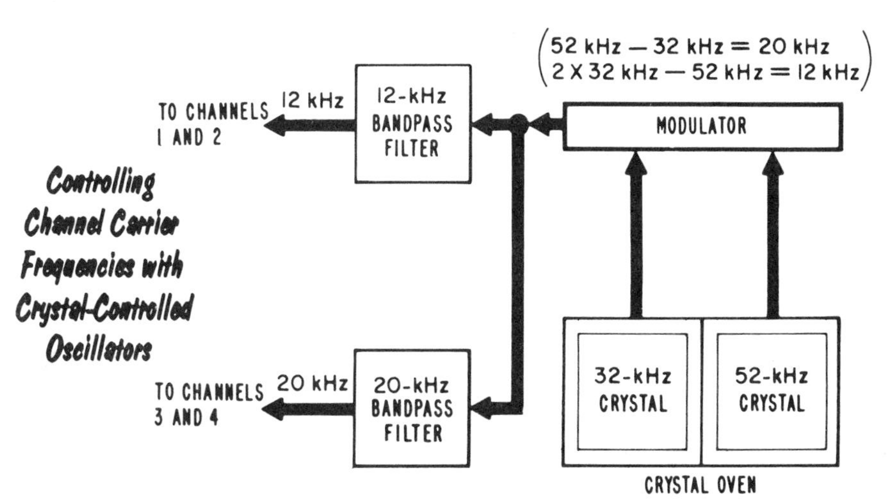

Controlling Channel Carrier Frequencies with Crystal-Controlled Oscillators

Carrier Frequency Control (cont.)

ceiving terminal as 220-3,420 Hz, assuming upper-sideband transmission. It is apparent, therefore, that the two oscillators, one that drives the modulator in the West transmitting terminal and the other driving the demodulation in the East receiving terminal, must be of the same frequency for proper system operation. This can be generally accomplished by synchronization of channel oscillators, pilot-frequency transmissions for control purposes, and use of one or two highly accurate temperature-controlled crystal oscillators at each carrier terminal to derive the channel carrier frequencies.

The channel oscillators may be aligned by checking with a 1,000-Hz voice frequency oscillator test set. After this initial alignment, periodic synchronization tests are made to insure alignments.

In most of the present carrier systems, continuous synchronization is provided by pilot-frequency transmissions. This makes use of separate continuous frequencies not related to the channel carrier frequencies. These signals are sent from one carrier terminal to the other one. The pilot frequency locks or synchronizes the corresponding carrier channel oscillators in the respective carrier terminals.

The diagram illustrates in simplified block form the use of two crystal-controlled oscillators for controlling channel carrier frequencies. With this arrangement, the frequency stability of the channel carriers can be maintained within about 2 to 6 Hz.

Questions and Problems

1. Name the principal types of carrier terminals and explain their differences.
2. What are the two main divisions of a carrier terminal?
3. What kind of carrier system is the C-type? Does it require two pairs?
4. What is the difference between the voice-frequency channel and a carrier channel such as in the C-type system?
5. What equipment is required to convert a four-wire carrier terminal to the two-wire type?
6. What relationships exist between (a) the directional filters at a carrier terminal; (b) the directional filters at the west and east carrier terminals?
7. Name the principal components of a four-wire carrier repeater for each direction of transmission.
8. What is the difference between a two-wire and a four-wire carrier repeater?
9. What means may be used for carrier frequency control in carrier suppressed type systems?
10. How may the frequency stability of channel carriers be maintained to within a few hertz?

Carrier Transmitted Systems

The early type C carrier system just studied is of the single sideband carrier suppressed type. Such systems require carrier reinsertion for demodulation purposes in the receiving branch and thus precise control and synchronization of channel carrier frequencies at carrier terminals. By transmitting the channel carriers, these requirements for satisfactory demodulation at the receiving terminal will not be necessary. The carrier transmitted method, which provides some equipment economies and less adjustment work, is used in several systems such as the Western Electric N1 and N2, and the GTE Lenkurt 47A/1 and 47A/2 double sideband cable carrier systems.

Let us examine the difference in basic system operations between the carrier transmitted and carrier suppressed modes. The drawing illustrates in simplified form one channel of a 12-channel group of the carrier transmitted system. In the carrier transmitted channel, (1) the carrier frequency is not suppressed in the modulator, (2) there is no need to reinsert carrier in the receiving branch for demodulation purposes, and (3) the filter requirements are not as severe. In the carrier suppressed channel, (1) the carrier is effectively suppressed in the balanced-bridge modulator, (2) the same channel carrier frequency is used for modulation and demodulation, and (3) an adequate filter is required in the transmitting branch to filter out the upper sideband.

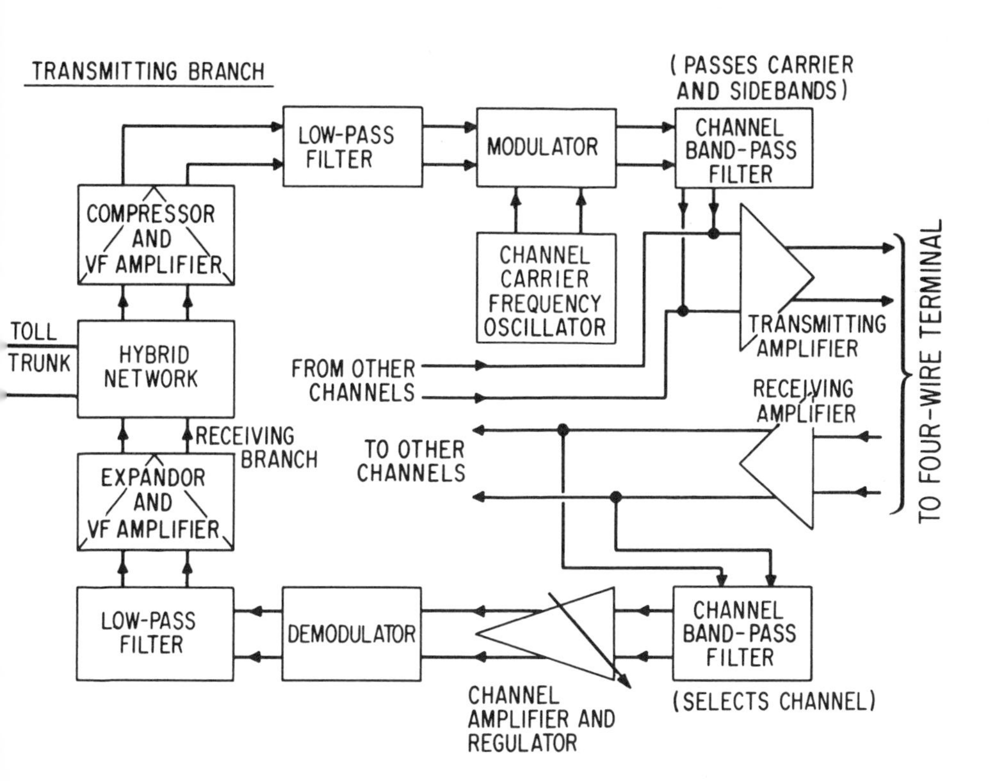

Single Sideband Transmission

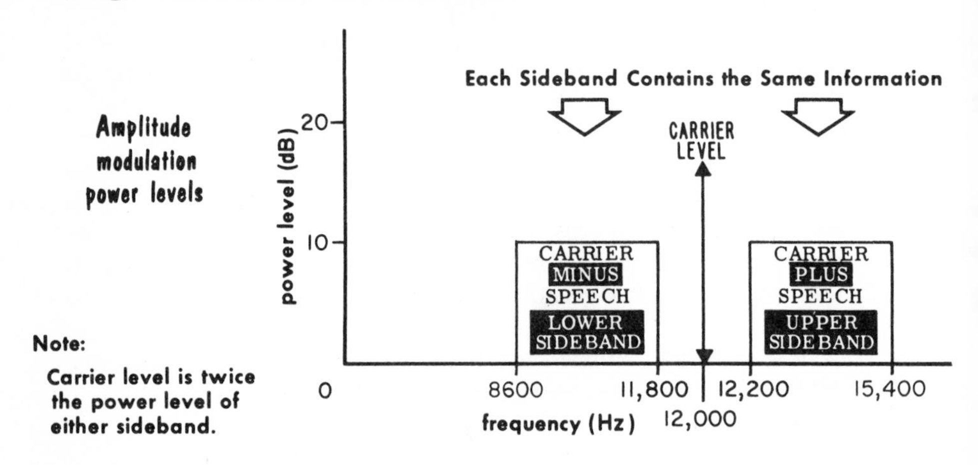

In the FDM Modulation and Demodulation Process section, we learned that these translation processes produced upper- and lower-sideband frequency components and that each sideband contains the original speech signals. In other words, both sidebands are carrying the same information. Consequently, only one sideband need be transmitted to convey speech intelligence from one carrier terminal to another. This arrangement, which can be achieved by suitable band-pass filters, is termed *single sideband transmission.* The carrier may be either transmitted or suppressed as determined by the design of the particular carrier system.

There are a number of advantages in transmitting only one side-band. A single sideband (SSB) system requires only one-half the bandwidth of a double sideband (DSB) system. We can therefore obtain twice the number of voice channels with SSB as we can with DSB. The SSB method will provide frequency economy, moreover, in carrier frequency allocations. Adjacent carrier channels can also be spaced closer together, resulting in a marked reduction in the over-all bandwidth of the carrier system. The bandpass filters needed for SSB transmission, however, are considerably more expensive than those required for DSB transmission.

For instance, double sideband carrier transmitted cable carrier systems, such as the GTE Lenkurt 47A/2, have a capacity of 12 voice channels. They utilize carrier frequency bands of 36 to 140 kHz and 164 to 268 kHz to transmit these 12 channels. Typical single sideband suppressed carrier systems, such as the GTE Lenkurt 46B, have a capacity of 24 voice channels. The carrier frequency bands used are 36 to 132 kHz and 172 to 268 kHz, for the respective low and high groups. Each group handles 24 VF channels in each direction.

Series-Resistance Pad Attenuation

The block diagrams of the carrier terminals showed pads in the voice frequency and carrier channels. These attenuation or resistance pads are often inserted in the receiving branch to decrease the strength of incoming carrier signals prior to demodulation. They also may be employed to decrease the level of speech signals before modulation or amplification. In this manner, pads prevent amplifiers from being overloaded. By installing a resistance pad at the input of a fixed-gain amplifier, it is possible to vary the overall gain. Pads can also be utilized under certain conditions to improve impedance matching. When used in conjunction with a voltmeter, pads can measure the gain of an amplifier.

Series Resistance Pad -- reduces signal strength to load

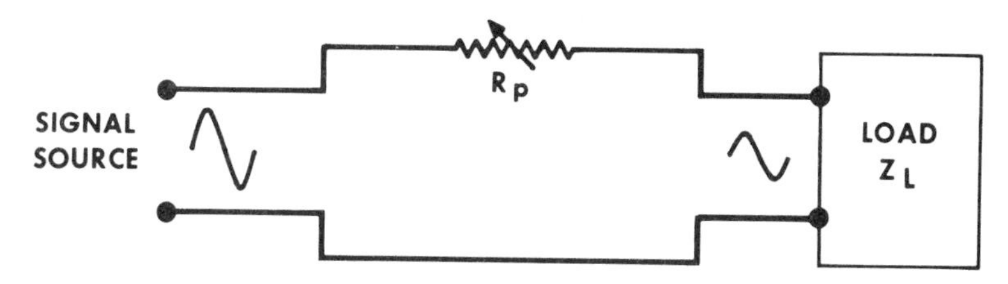

SERIES RESISTANCE METHOD OF ATTENUATION

The schematic illustrates one of the two basic methods of attenuation by the use of resistors. The two basic means are the series-resistance and the shunt-resistance arrangements. In the series-resistance arrangement, the resistance pad, R_p, reduces the signal strength of the signal to the load for the following reasons:

The voltages across the elements of a series circuit are directly proportional to the resistance of the elements. Power in the circuit is equal to E^2/R.

Therefore, the voltage across the load, Z_L, will be less with R_p in the circuit. Since the voltage across the load has been reduced, the power or signal strength delivered to the load, Z_L, also is decreased. Consequently, by inserting a series resistor, the transmitted signal can be attenuated. The larger the value of this series resistor, the greater will be the attenuation of the signal.

Shunt-Resistance Attenuation Method

The shunt-resistance method is the other basic means of attenuating signal levels in carrier channels and terminal equipment. In the diagram the shunt resistance or pad, R_p, is in parallel with the voltage source and the load. Therefore, the total resistance across the signal or voltage source is decreased. This causes a larger current to flow from the signal source. The resultant larger voltage drop in the circuit reduces the voltage across the load, Z_L. A decrease in voltage across a resistor also causes less power to be delivered to it since $P = E^2/R$. Therefore, the signal power delivered to the load is less. In this case, the smaller the shunt resistance, R_p, the greater the attenuation.

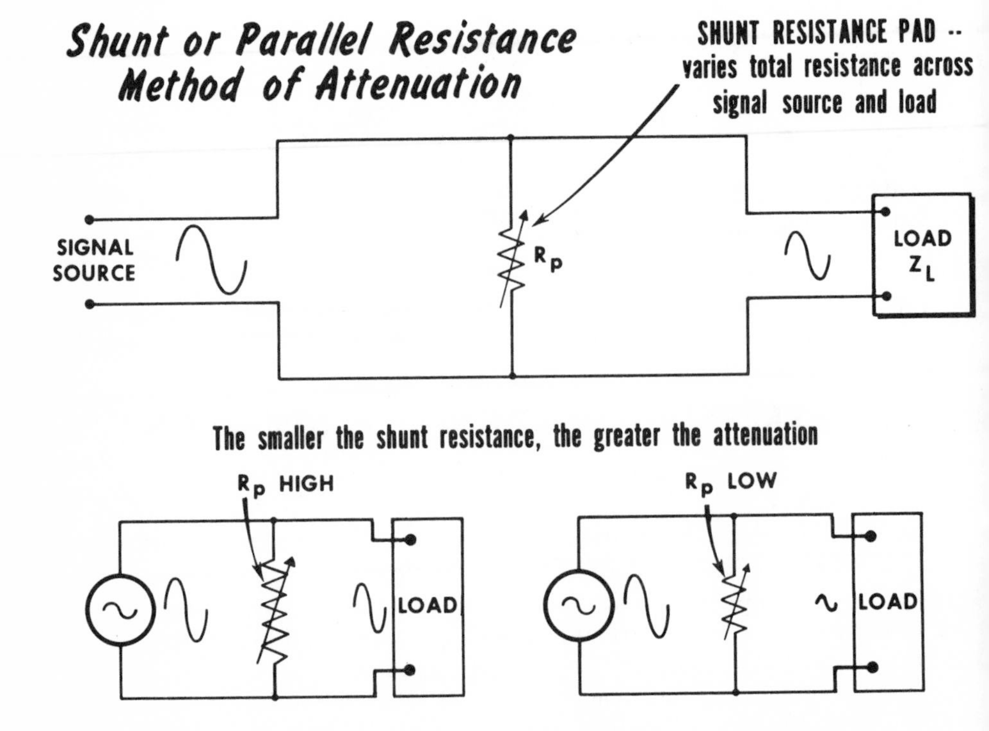

The use of either series or shunt resistors to attenuate signals is not satisfactory for telephone transmission. These basic arrangements result in mismatched networks which cause reflection of signals and harmful echo effects. Is is essential that attenuation pads comprise symmetrical arrangements of resistors to maintain the impedance match between networks, especially in carrier terminals. This is accomplished by using combinations of series and parallel, or shunt resistors. The following pages will discuss these symmetrical pads in more detail.

Types of Symmetrical Pads—T Pad

There are four types of attenuation pads generally used in voice frequency and carrier circuits. They are the: T, Pi, H, and square pad. The T and Pi type pads can be symmetrically designed. However, these pads are not balanced to ground and their use is restricted. They may be used, for example, when the side of the line not containing the series resistor is grounded. These pads also can be used when it is unnecessary to have the attached networks balanced to ground.

The symmetrical T pad is inserted between two matched networks, as, for example, between two 600-ohm terminations. This pad consists of two equal series resistors designated R1. One is placed on each side of shunt resistor R2. The design of a T pad can be computed from design equations or from a table derived from the design equations. It is necessary to know the desired attenuation and the impedance value of the circuit into which the T pad is to be inserted. As an illustration of typical design values, let us assume that an attenuation of 20 dB is desired from a symmetrical T pad, to be inserted in a 600-ohm circuit. From the design tables we obtain K factor values of 0.8182 and 0.2020 for the series and shunt arms, respectively. Since the circuit impedance is 600 ohms, the required resistance is: R1 = 600 × 0.8182 = 490.9 ohms; R2 = 600 × 0.2020 = 121 ohms.

SYMMETRICAL T PAD

R1 = series arm resistors

R2 = shunt arm resistor

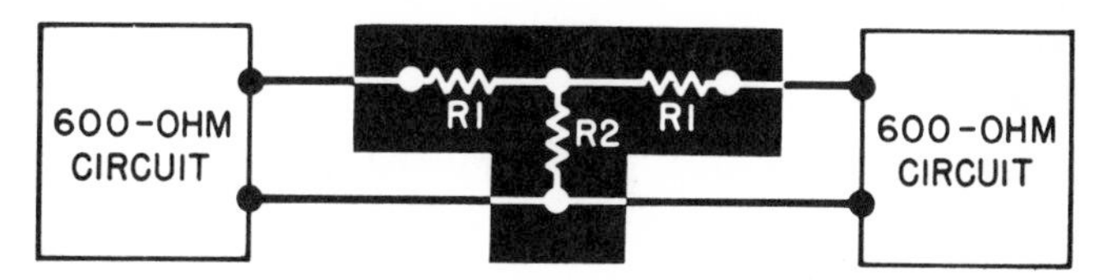

Shunt and Series Arm Factors for T and H Pads between Circuits of Equal Impedance

dB	*Series-arm factor*	*Shunt-arm factor*	*dB*	*Series-arm factor*	*Shunt-arm factor*
1	0.0575	8.6709	21	0.8363	0.1797
2	0.1146	4.3045	22	0.8528	0.1599
3	0.1710	2.8384	23	0.8678	0.1423
4	0.2262	2.0966	24	0.8813	0.1267
5	0.2802	1.6448	25	0.8935	0.1128
6	0.3323	1.3386	26	0.9046	0.1005
7	0.3825	1.1160	27	0.9145	0.0901
8	0.4305	0.9462	28	0.9234	0.0797
9	0.4762	0.8118	29	0.9315	0.0711
10	0.5195	0.7028	30	0.9387	0.0633
11	0.5603	0.6123	31	0.9452	0.0564
12	0.5985	0.5362	32	0.9510	0.0503
13	0.6342	0.4714	33	0.9562	0.0448
14	0.6673	0.4156	34	0.9609	0.0399
15	0.6980	0.3673	35	0.9651	0.0356
16	0.7264	0.3251	36	0.9688	0.0317
17	0.7525	0.2883	37	0.9721	0.0283
18	0.7764	0.2558	38	0.9751	0.0252
19	0.7982	0.2273	39	0.9778	0.0224
20	0.8182	0.2020	40	0.9802	0.0200

Symmetrical Pi Pad Design

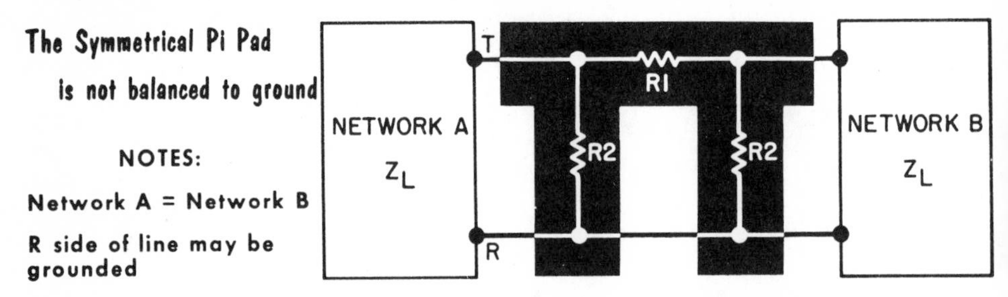

The Pi pad may be inserted between two matched networks when one side of these networks (the side not containing the series resistor of the Pi pad) is grounded. Similarly, the Pi pad may be used when the attached circuit or networks do not have to be balanced to ground. The design of Pi pads may be calculated directly from design equations or by a table derived from the design equations. It is necessary, as in the case of T pads, to know the desired pad attenuation in decibels and the resistance value of the circuit into which the Pi pad will be inserted. The following general equations may be used for Pi pad design:

$$\text{Pad loss in decibels} = 20 \log K \tag{1}$$

$$\text{Series-arm resistance R1} = R\,\frac{K^2 - 1}{2K} \text{ ohms} \tag{2}$$

$$\text{Each shunt-arm resistance R2} = R\,\frac{K + 1}{K - 1} \text{ ohms} \tag{3}$$

Where R = resistance value of circuit into which Pi pad is to be inserted, and K = factor obtained from Equation (1) for the desired pad loss to be inserted.

To illustrate the application of the above formulas, let us design a Pi pad to give a 20-dB loss in a 600-ohm circuit. From Equation (1) the value of factor K is given by: 20 = 20 log K or 1 = log K. Since the antilog of 1 is 10, K = 10. Therefore, by substitutions in Equations (2) and (3) above, we obtain the following values for the series- and shunt-arm resistors:

$$R1 = 600\,\frac{(10)^2 - 1}{2 \times 10} = 2{,}970 \text{ ohms for the series arm}$$

$$R2 = 600\,\frac{10 + 1}{10 - 1} = 733.3 \text{ ohms for each resistor in the shunt arm}$$

A table of shunt and series arm factors for Pi pads can be constructed from the above design equations. This table would show for each desired attenuation value, say from 1 to 40 dB, the series- and the shunt-arm factors. Multiplying the known circuit or network resistance by these factors expedites Pi pad design calculations.

Balanced Pads—H Type

The H and square type pads are symmetrical and balanced to ground, used only with networks balanced to ground, and never used in circuits where one side of the line is grounded.

A typical symmetrical H pad is derived directly from the T pad previously described. In fact, an H pad may be made from a T pad in the following manner: divide each series resistor R1, of the T pad into two equal resistors. Place one resistor, ½ R1, in the opposite line conductor on the same side of R2. Do the same on the other side of R2. The result is the H pad.

The design equations of the H pad are also derived directly from the T pad. These equations are as follows:

$$R1 = R\,\frac{K-1}{K+1} \quad \text{or} \quad \frac{R1}{2} = \frac{R\,(K-1)}{2\,(K+1)} \text{ ohms} \tag{4}$$

$$R2 = R\,\frac{2K}{K^2-1} \text{ ohms} \tag{5}$$

where the value of K is as given by Equation (1).

The above equations also may be written as follows for use with the design table:

$$\frac{R1}{2} = \frac{R}{2} \times \text{series-arm factor from the table} \tag{6}$$

$$R2 = R \times \text{shunt-arm factor from the table} \tag{7}$$

To illustrate the design of a typical H pad, assume that a 10-dB loss is desired for insertion in a 500-ohm circuit. From the T and H design table, it is ascertained that for a 10-dB loss, the series-arm factor is 0.5195 and that the shunt-arm factor is 0.7028. Therefore, by substituting in Equations (6) and (7) the following values are obtained:

$$\frac{R1}{2} = \frac{500 \times 0.5195}{2} = 129.88 \text{ ohms}$$

$$R2 = 500 \times 0.7028 = 351.4 \text{ ohms}$$

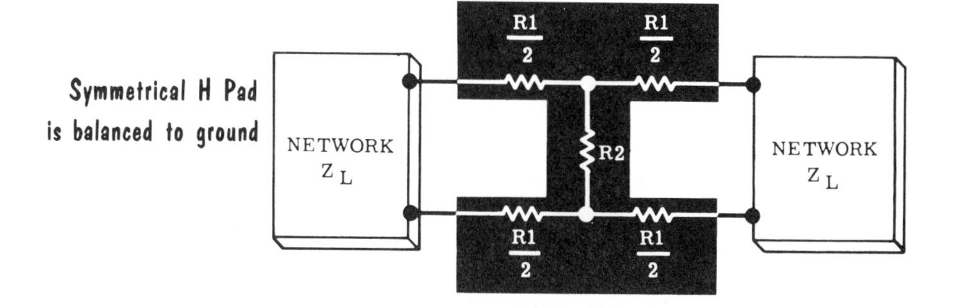

Square Pad Design

The symmetrical square pad is derived from the symmetrical Pi pad described previously. In a similar manner as the H pad was made from the T pad, the square pad can be formed from the Pi pad. The series-arm resistance of the Pi pad can be divided into two equal resistors to form the square pad. The schematic diagram illustrates this.

SYMMETRICAL SQUARE PAD BALANCED TO GROUND

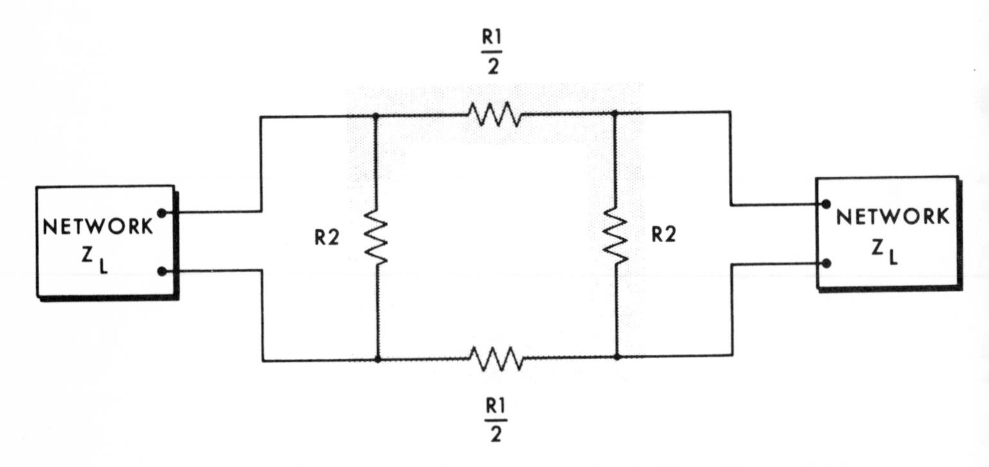

Square Pad can be formed from Pi Pad with series-arm resistance of Pi pad divided into two equal resistors.

The design equations of the symmetrical square pad, like the pad itself, are derived directly from the Pi pad. The square pad can be designed by the direct application of Equations (1)—(3) as given previously on page 118. Square pads also can be designed by a table which is derived from these aforementioned equations.

Let us refer to the design example which was given for the Pi pad as an illustration of designing square pads. In this aforementioned example, a 20-dB loss was desired from a pad to be inserted in a 600-ohm circuit. Therefore, for a square pad, the values are as follows:

$$\frac{R1}{2} = \frac{R\,(K^2 - 1)}{2\,(2K)} = \frac{600\,(10^2 - 1)}{2\,(2 \times 10)}$$ 1,485 ohms for each resistor in the series arm

$$R2 = \frac{R\,(K + 1)}{K - 1} = \frac{600\,(10 + 1)}{10 - 1}$$ 733.3 ohms for each resistor in the shunt arm

Use of Equalizers

In the section on Telephone Carrier Elements, we learned that an *equalizer* adds attenuation to a circuit. The purpose is that the total loss or attenuation will be about the same for all frequencies transmitted. This requirement stems from the characteristics of telephone transmission lines. For example, frequencies in the range of 1,000 to 3,400 Hz are attenuated more than the low frequencies in the 200- to 1,000-Hz range. This effect causes distortion to the speech signals if not corrected. The attenuation equalizer can correct for this condition. The attenuation for any frequency is directly proportional to the line length. Thus as the line transmission length increases, the amount of distortion becomes greater. By using a properly designed equalizer, the attenuation can be made the same for all frequencies transmitted. Equalizers are usually installed at the receiving end of carrier systems.

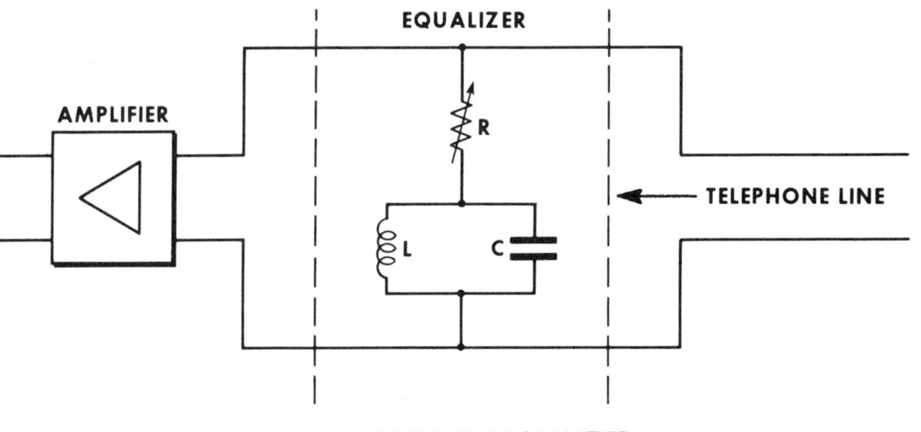

ATTENUATION EQUALIZER

Provides Attenuation to Equalize Voice Frequencies Transmitted

The schematic diagram shows an equalizer installed before the receiving branch amplifier of a carrier terminal. It consists of an inductor L, capacitor C, and variable resistor R. Like filters, the design of an equalizer can be very complicated. The attenuation of an ideal equalizer should decrease with an increase in frequency. This means that the impedance of the equalizer has to increase with an increase in frequency. This is accomplished by the parallel resonant circuit of L and C. The amount of current flowing through R decreases as the frequency increases. Since power equals I^2R, the amount of power dissipated or lost in the equalizer decreases with an increase in frequency. By varying R, the amount of this dissipated power can be changed. This permits the equalizer to be adjusted to obtain the correct attenuation for different line lengths or conditions.

Carrier-Signaling Methods

In the early days of carrier systems, signaling over toll circuits was normally accomplished on a ringdown basis. The standard 20-Hz ringing voltage was changed to 1,000 Hz by a 20/1,000-Hz voice frequency ringer associated with each carrier channel. The 1,000-cycle tone was interrupted at a rate of 20 cycles per second. The resultant voice frequency ringing signal was then transmitted over the carrier channel facilities to the distant carrier terminal. At the far end, the 1,000-Hz signals operated relay equipment in the toll trunk circuit, causing a local source of 20-Hz ringing current to be connected to the switchboard. The combination of the 1,000-Hz signal interrupted at the 20-Hz rate was used to prevent possible false operations of the relay equipment from speech signals during conversation.

The ringdown signaling method was satisfactory as long as toll and long distance calls were handled manually by operators. The advent of *operator toll dialing* and then *direct distance dialing* (DDD) over the nationwide automated intertoll network made ringdown operations obsolete. The *single frequency* (SF) and *multifrequency* (MF) signaling systems were subsequently developed, particularly for use over carrier circuits. The two forms of SF signaling now in use are defined as *inband* and *out-of-band.* In both systems, the signaling, supervisory, and dialing dc signals of the intertoll trunk circuits are transformed to a-f tones for transmission between carrier terminals. At the receiving carrier terminal, these audio signals must be converted back to dc signals as originally initiated because a dc path does not exist for carrier channels in a carrier system.

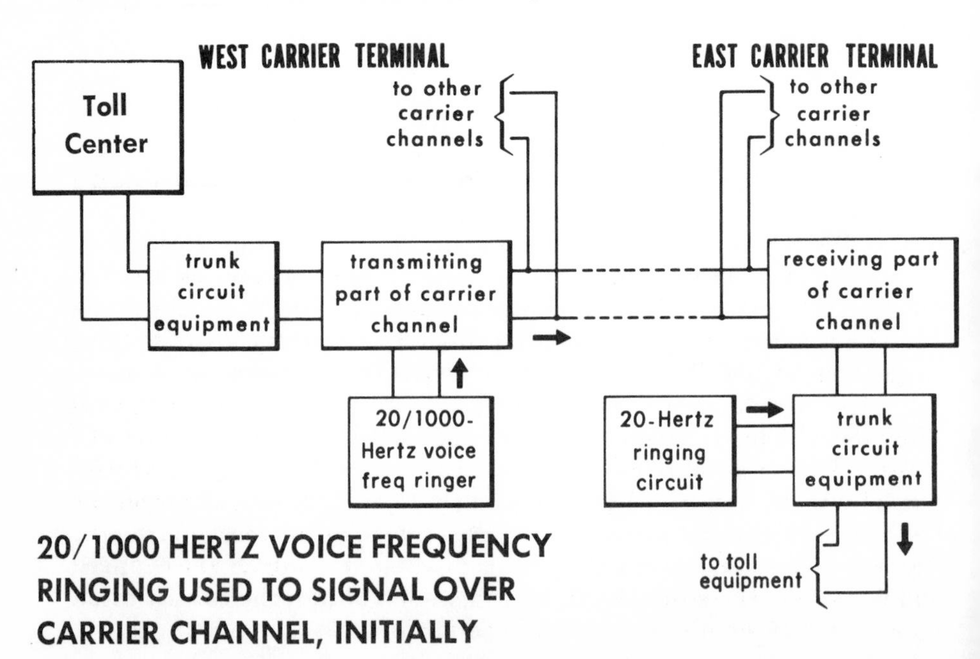

20/1000 HERTZ VOICE FREQUENCY RINGING USED TO SIGNAL OVER CARRIER CHANNEL, INITIALLY

Inband Signaling Systems

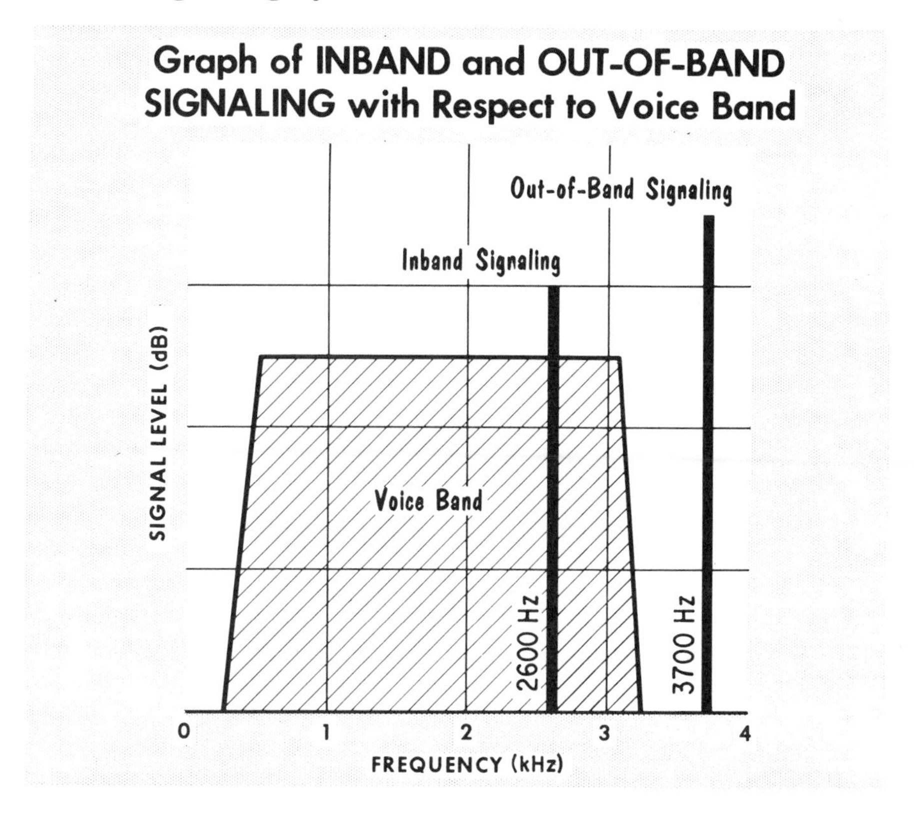

Signaling systems in telephone service that make use of frequencies within the transmitted voice band (200-3,400 Hz) are known as *inband signaling systems.* Both the voice and signaling paths are the same in such systems. Therefore, special equipment is not required to coordinate the two types of signals. They are transmitted over the same voice path. The equipment required for handling inband signaling is needed only at the end terminals. Two or more carrier systems employing inband signaling may be operated in tandem without additional intervening signaling apparatus. A disadvantage of inband signaling is the possibility of false operation of the signaling equipment during the transmission of speech.

The present standard frequency for single-frequency (SF) inband signaling is 2,600 Hz. It is utilized in carrier systems for toll-connecting and intertoll trunks, such as the Western Electric N2, N3, the GTE Lenkurt 47A/N2, 46B, and similar types. When a carrier system employs the same wire pair for two-way transmission, 2,600 Hz is used for signaling in one direction and 2,400 Hz for signaling in the opposite direction. The diagram shows the relationships between the voice band, the inband, and the out-of-band signaling frequencies.

Inband Signaling Operations

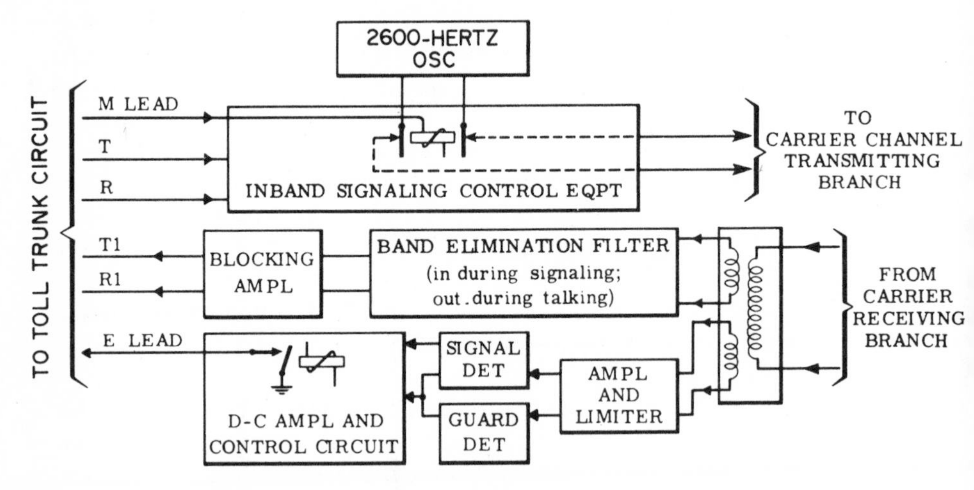

2600-HERTZ SINGLE-FREQUENCY INBAND SIGNALING CIRCUIT

To better understand signaling over a carrier channel, we should know the significance of the E and M leads, discussed in Telephone Transmission Practices. In essence, these two leads serve as the communication link between the toll trunk circuit and the particular signaling system used with a carrier channel.

The M lead transmits the near-end trunk supervisory signals to the carrier channel for transmission to the distant end. The E lead receives from the signaling system of the carrier channel the indication of the supervisory conditions at the far end. For example, an idle trunk or *on-hook* signal is on the M lead. It causes the 2,600-Hz tone to be applied to the transmitting branch of the carrier channel. This may be termed an idle trunk condition. When the trunk is busy, an *off-hook* signal is placed on the M lead. This causes the 2,600-Hz tone to be removed from the carrier channel. Similarly, the presence or absence of incoming 2,600-Hz signaling tones in the receiving branch of the carrier channel actuates its signaling system. It applies corresponding periods of pulses on the E lead, indicating *on-hook* and *off-hook* conditions, respectively, existing at the distant end. Dial pulses can be sent similarly. Safeguards against false operations during speech are necessary. Signal and guard detector devices are used. The signal detector operates on the 2,600-Hz tone; the guard detector on other speech frequencies. The outputs of the two detectors oppose each other and provide additional margin against false operation during conversation. The additional equipment for safeguarding inband signaling systems against interference from speech signals increases the maintenance problems and the costs. Another disadvantage is that the speech band is slightly degraded because of the slot (about 60-Hz wide) taken at 2,600 Hz for inband signaling.

Out-of-Band Signaling System

Frequencies just above the voice band, normally in the 3,400-4,000-Hz range, are employed for *out-of-band* signaling systems. The standard 20-Hz ringing current may also be considered as an out-of-band signaling method. This low frequency, however, cannot be handled by carrier and voice-frequency equipment. Therefore, it is not used for signaling purposes. The main advantages of out-of-band signaling are relative freedom from speech interference, and the use of higher signal levels to improve signaling reliability. There is a disadvantage, however, such as when it is necessary to interconnect the voice channels of two separate carrier channels. In such cases, the signaling leads of each voice channel must be connected through additional signal converter equipment.

The out-of-band signaling frequency, commonly employed for toll-connecting and intertoll trunks in the DDD network, is 3,700 Hz. In Europe and other countries, the 3,825-Hz-frequency is usually provided in accordance with CCITT standards. The resultant useful voice band is somewhat narrower than that obtained with inband signaling, because of the filters needed to protect against the higher level 3,700-Hz signals. Carrier systems equipped for 3,700-Hz out-of-band signaling include the Western Electric N1, N2, O, and ON types; and the GTE Lenkurt 47A/N1.

The illustration shows a simplified block diagram for a 3,700-Hz out-of-band signaling system. The 3,700-Hz tone is transmitted during the trunk idle or on-hook state. No tone is sent during the busy or off-hook mode. A varistor or other electronic-type keyer, under control of the M lead, applies a 3,700-Hz tone to the channel modulator in the transmitting branch of a channel carrier unit. At the distant end, this tone is selected at the channel demodulator output by a 3,700-Hz

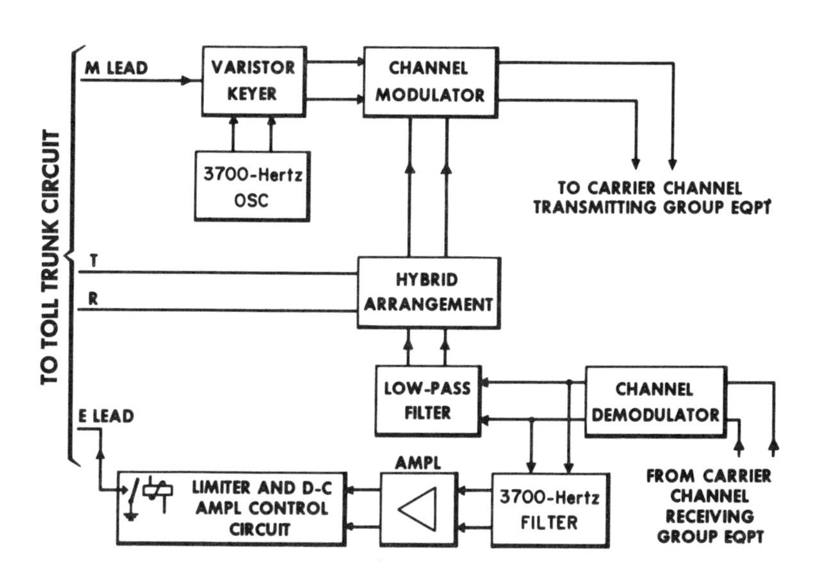

Out-of-Band Signaling System (cont.)

filter and is amplified and rectified. The resultant direct current controls set a relay or an electronic device to establish the desired E lead condition.

Common Channel Interoffice Signaling Concept

As previously explained, both inband and out-of-band signaling systems tend to slightly degrade their associated voice circuits. Moreover, the additional apparatus and filters, which are needed to prevent interference to speech signals, increase the cost of the channel carrier equipment. In addition, SF inband signaling systems are subject to *talk-down* and *talk-off* conditions, as well as unauthorized tone signals that might originate at customers' stations. The use of carrier systems for interoffice and toll-connecting trunks has increased greatly in recent years, leading to the development of the *Common Channel Interoffice Signaling* (CCIS) concept by the Bell System, and to the *Common Channel Signaling Systems* (CCSS) used by non-Bell System telephone companies.

The common channel signaling method may be defined as a process for utilizing a single voice-band channel for time-division assignments of up to 24 signaling circuits, associated with a cable carrier system, for transmission to the distant carrier receiving terminal. A separate voice channel is employed for each direction of transmission. At the distant carrier terminal, the 24 signaling circuits are re-assigned in the same time sequence as their respective carrier channels. The CCIS process affords a more economical and reliable means for conveying signaling information over cable carrier systems than conventional SF inband signaling

In the common channel signaling process, the E&M leads (refer to pages 53 and 124) of each trunk circuit are connected to the common signaling equipment. The M lead condition of the trunk circuit is translated into logic levels for conversion into binary digits or bits (zeros and ones). The resultant data stream of binary digits from the 24 trunk circuits is then time-division multiplexed (TDM) over a designated voice channel, usually at the rate of 2,400 bits per second. At the distant carrier terminal the received data is decoded, demultiplexed, and distributed to the respective 24 trunk signaling channels in the same time sequence as originally transmitted. The overall procedure may be likened to the use of Dataphone Sets for sending information between computers over telephone circuits.

To understand the common channel signaling technique, we will consider the operation of a typical common channel signaling system, such as the GTE Lenkurt 11A CCSS. The simplified block diagram shows the major elements of this particular signaling system usually associated with a standard 24-channel cable carrier system.

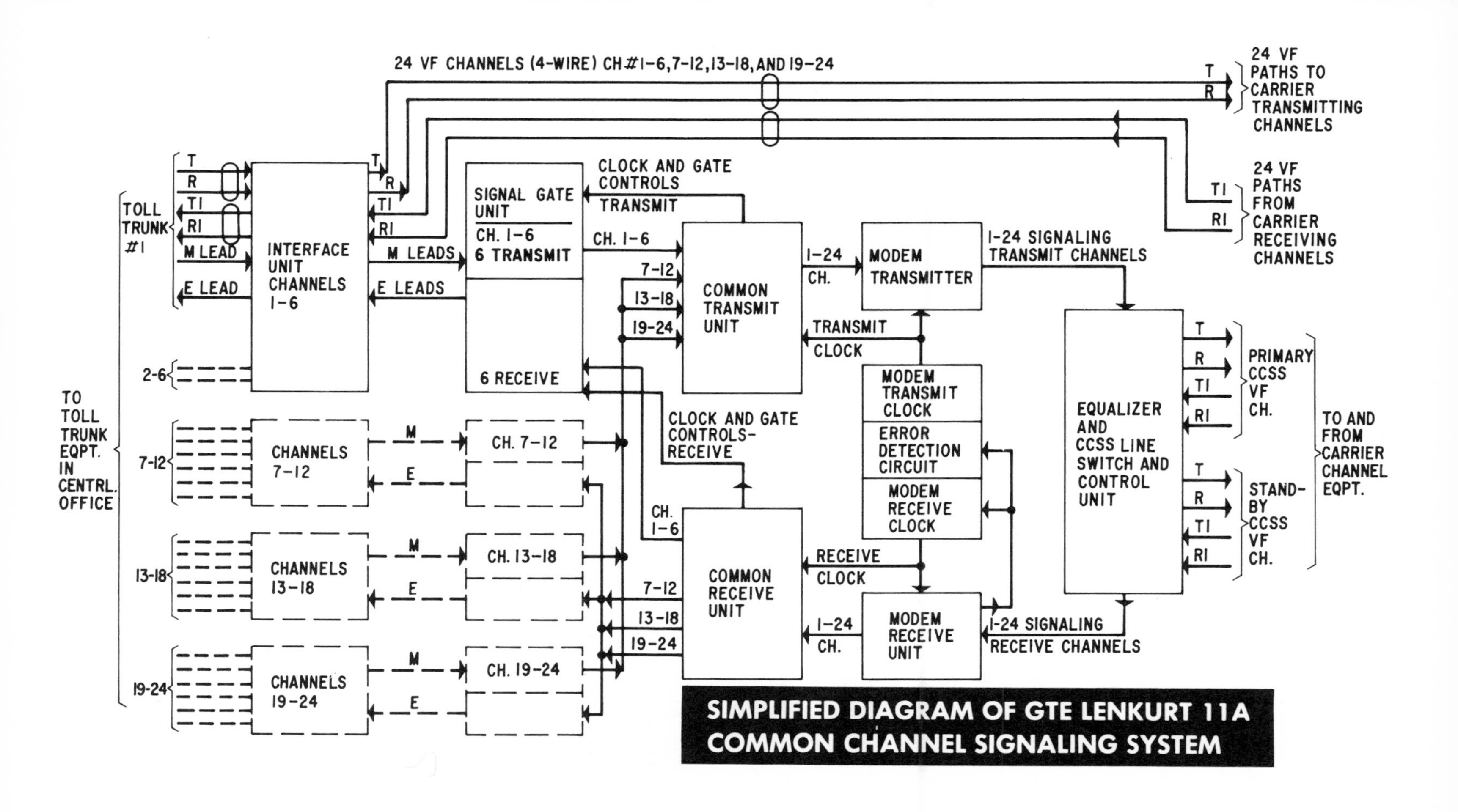

SIMPLIFIED DIAGRAM OF GTE LENKURT 11A COMMON CHANNEL SIGNALING SYSTEM

Common Channel Interoffice Signaling Concept (cont.)

The *interface units* provide the interconnections between this CCSS and the central office or toll center trunk equipment. The 4-wire transmission paths (T,R, and T1,R1) of the 24 voice circuits are routed directly to their respective carrier channels. The E&M signaling leads (discussed in Inband Signaling Operations) of each trunk circuit connect to their respective interface unit. The signaling state on the particular trunk's M lead is converted into logic signals by the interface unit. For instance, ground on the M lead (indicating an idle trunk or on-hook state) is converted to -5 volts for a logic signal of binary digit 1. Battery potential on the M lead (indicating a busy trunk or off-hook state) would be converted to 0 volts (ground potential) for a binary digit 0 logic signal. Four interface units are furnished, each with a capacity of six trunk circuits; these provide signaling for up to 24 trunk carrier circuits.

The binary logic signals (zeros and ones) are sent to the *transmit gates* in the four *signal gate units.* Each signal gate unit handles the binary logic signals from six interface units, for a total of 24 signaling channels. The signal gate unit conditions the transmit logic levels to guard against false signals caused by impulse noise or other interference. Each signal gate unit also corrects and shapes the binary pulses, from its six associated interface units, before sending them to the *common transmit unit.* The binary pulses from the four signal gate units (total of 24 signaling circuits as previously described) are sequentially sampled by the common transmit unit to produce binary signal samples at a rate of 2,400 samples or bits per second. These samples are controlled and synchronized by the *modem transmit clock.* The resultant sample stream of binary pulses is next presented to the *modem transmitter* in the modem logic unit.

The *moden transmitter, modem transmit clock, modem receive clock,* and the *error detection circuits* are contained in the modem logic unit. The stream of binary pulses (from 24 signaling channels) received in the common transmit unit is then modulated to a *frequency-shift keying* (FSK) type of signal by the modem transmitter. The resultant FSK binary signal occupies a 600-Hz to 3,000-Hz bandwidth. It is routed to the primary voice channel assigned to the CCSS for transmission to the far-end carrier terminal.

At the far-end carrier terminal, the incoming signals are demodulated in the *modem receiving unit,* shown in the block diagram. The demodulated signals are also conveyed to the *error detection circuit* for a check of any errors in the code pattern. An excessive error rate, loss of carrier, or degradation of the signal level will initiate the transfer of the primary voice channel to a stand-by channel. The modem receive unit also regenerates and retimes the received FSK binary pulses into a train of binary pulses that represent the original signaling states of the 24 trunk circuits. The *modem receive clock* retimes

Common Channel Interoffice Signaling Concept (cont.)

the received pulse train for presentation to the *common receive unit.*

The aforementioned pulse train is decoded within the common receive unit, which likewise generates 24-channel pulses for routing the received binary digits (bits) to the proper signal gate unit. The common receive unit also distributes the received binary logic signals to the respective receive section of the signal gate unit for the six circuits it is associated with. Each receive signal gate unit furnishes the necessary logic levels to its associated interface units. In the interface unit, the binary digits (zeros and ones) representing the logic levels received from the signal gate unit are converted into dc signaling states. For example, a binary digit 0 indicates the on-hook state which produces an open-circuit on the E lead. A binary digit 1 indicates the off-hook state, which results in a ground on the E lead.

Multifrequency Pulsing System

Multifrequency pulsing (MF) was developed by the Bell System for conveying information between dial central offices and among toll switching centers. This system transmits each digit of a number or other address information with a single spurt of audio-frequency tones within the 700-1,700 Hz band. It is adaptable for transmission over any voice channel (200-3,400 Hz). MF also has the advantage of providing high-speed signaling over intertoll facilties and to transmit operator key-pulsing signals from toll switchboards.

Each digit (0 to 9) comprises a sole combination of two out of five possible frequencies in the 700- to 1,500-Hz range. A sixth frequency (1,700 Hz) is used in combinations to provide additional signals for control and other functions. The six frequencies are spaced 200 Hz apart in the 700- to 1,700-Hz region. Each combination of two fre-

MULTIFREQUENCY SIGNALING CODE

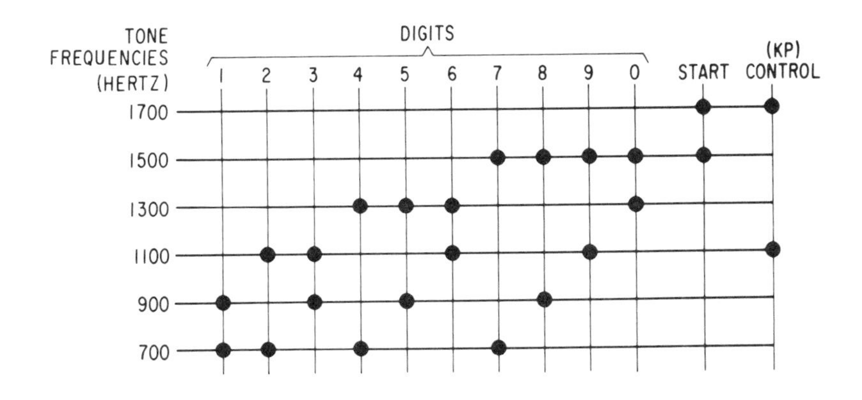

Multifrequency Pulsing System (cont.)

quencies represents a particular digit or pulse. The MF system is designed so that if more than two frequencies are detected for a particular digit, the receiving equipment will reject the signals and return a reorder indication.

The chart on p. 129 shows the digit codes and frequencies for the multifrequency signaling system. For example, digit 5 is represented by a signal composed of a mixture of 900-Hz and 1,300-Hz tones. Digit 0 consists of a signal depicted by a combination of 1,300- and 1,500-Hz frequencies. The start and control signals designated in the chart are used for supervision and control purposes in intertoll switching operations.

Questions and Problems

1. What are the main differences between a carrier suppressed and a carrier transmitted system?
2. What are the advantages of transmitting only one sideband?
3. What sideband is normally transmitted in a SSB carrier system? Why?
4. Name four general types of attenuation pads. Which ones are commonly used with balanced networks?
5. Does the attenuation of an equalizer increase or decrease as the frequency increases? Why?
6. What caused the obsolescence of ringdown signaling?
7. What two signaling systems are generally used on the intertoll network?
8. What particular frequency is currently employed for inband signaling in carrier systems?
9. What improved type of signaling system has been recently introduced for interoffice and toll cable carrier systems? What condition on the M lead indicates an idle trunk? An off-hook state?
10. What frequency range is utilized by the multifrequency pulsing system? What frequencies comprise the digit 2?

Demise of Open-Wire Lines

The initial carrier systems, such as the type C previously described, were designed for operation over open-wire telephone lines. The vast majority of toll and long-distance circuits, in the early days of carrier developments, were on pole lines. Moreover, by confining the previous explanations mainly to open-wire applications, it was possible to present carrier telephone fundamentals in a simplified manner. This method would also enable the reader to better understand the principles involved.

There have been many new developments and rapid growth in toll and long distance facilities during the past two decades. For instance, multiconductor and coaxial cables have replaced open-wire pole lines. Also, point-to-point microwave systems and satellite circuits currently provide the majority of long distance trunks within the United States and Canada. Likewise, improved semiconductor devices, particularly integrated circuits, have made it economical to employ carrier on cables for relatively short-distances, including suburban and rural subscriber areas. Carrier cable systems of the Pulse Code Modulation (PCM) type are being utilized more and more for interoffice and toll-connecting trunks. The chart on this page graphically shows the percentage change in the kind of transmission facilities employed for toll and long distance circuits by the Bell System, in the past decade.

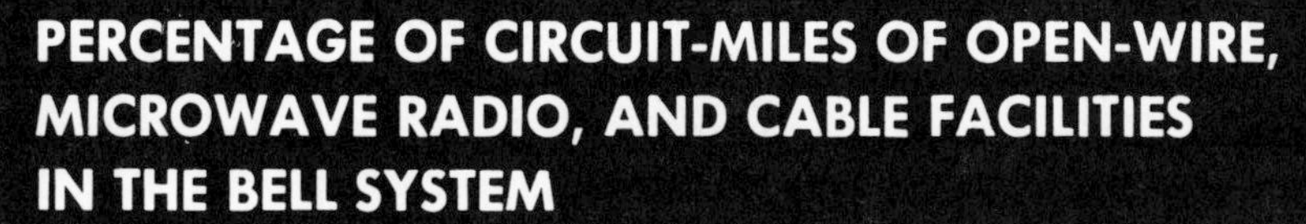

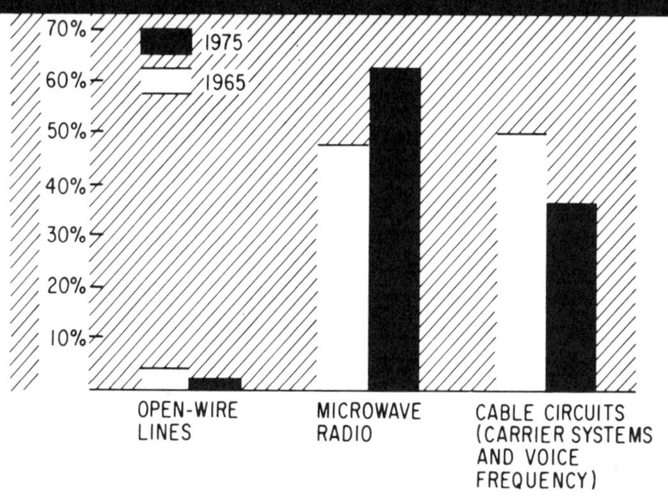

Cable Characteristics for Carrier Transmission

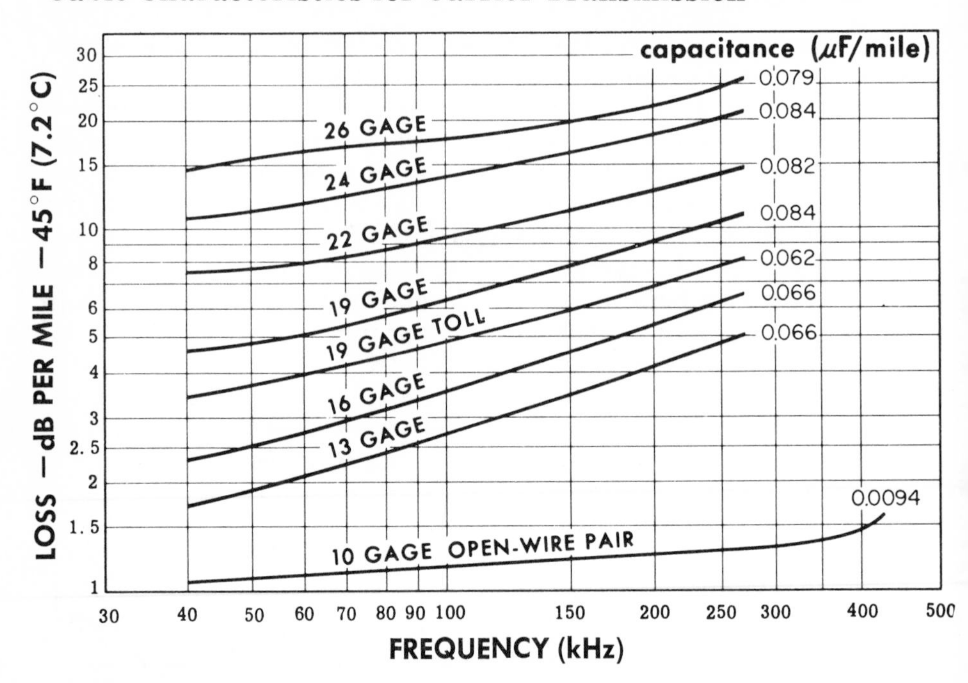

ATTENUATION VS. FREQUENCY CHARACTERISTICS OF CABLES

It was pointed out in Wire Transmission Principles that open-wire lines present more or less equal attenuation to carrier frequencies. This referred particularly to the 4,000-150,000-Hz range. The size of the wires used on pole lines has relatively little effect on its attenuation to carrier frequencies. This condition, however, does not apply to telephone cables. Changes in the wire size of cable pairs have a profound effect upon the attenuation to carrier frequencies. In general, attenuation increases directly with the length of the telephone circuit. This is the case for both open-wire lines and cable pairs. However, the attenuation of a cable pair rises rapidly as the carrier frequency increases.

The graph illustrates this attenuation effect. The main reason for this situation is the relative high capacitance existing between wires of a cable pair. The capacitance of a cable pair is very much greater than that of an open-wire pair. In a cable, the conductors of a pair are close together. The paper or plastic insulation which separates them has a high dielectric factor as compared to air. For example, the capacitance of a mile of No. 22 gage cable pair is 0.082 microfarad. Toll cable of No. 19 gage has a capacitance of about 0.062 microfarad per mile. On the other hand, the capacitance of a 104-mil open-wire line spaced 8 inches is only 0.0091 microfarad per loop mile.

Cable Impedance Effects

The inherent high capacitance of cable pairs causes other effects as the frequency increases. One important result is that the effective impedance of the cable pair is lowered. This is because the capacitive reactance, X_C, decreases as the frequency increases. This reduction in the impedance of a cable pair, however, is not linear with frequency. As the carrier frequency goes up, the inductive reactance, X_L, of the cable pair tends to increase. The result is that the impedance decreases quite rapidly through the lower range of carrier frequencies. As the carrier frequency continues to increase, the impedance tends to level off. This condition is shown in the graph.

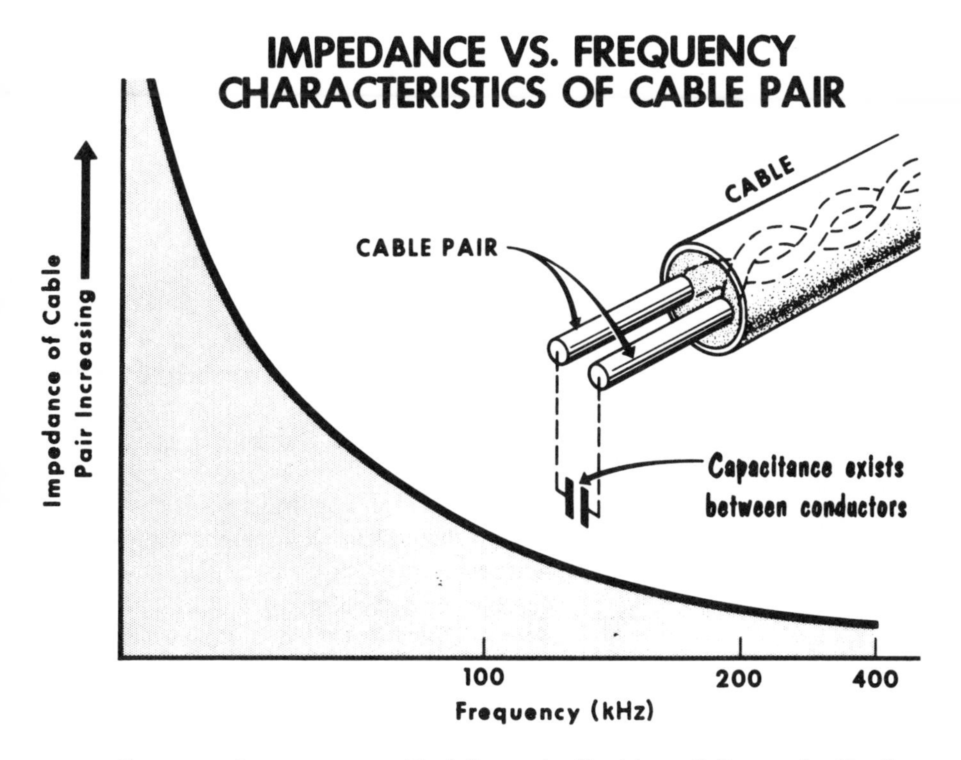

In general, we can say that the net effect is as follows: As the frequency increases, the capacitance effect of a cable pair offers more parallel paths to the flow of the carrier currents. This causes an apparent lowering in the impedance of the circuit. Therefore, more of the carrier current is lost in the cable and less power reaches the distant end. As an example, a No. 22 gage nonloaded cable has a loss of almost 3 dB per loop mile at 3,000 Hz. At 20,000 Hz this loss is 6 dB, at 60 kHz it is 8 dB and at 100 kHz the loss is about 9.5 dB. This high attenuation loss to carrier frequencies requires that carrier repeaters be employed at frequent intervals to insure reliable transmission of the carrier channels.

Slope and Temperature Effects

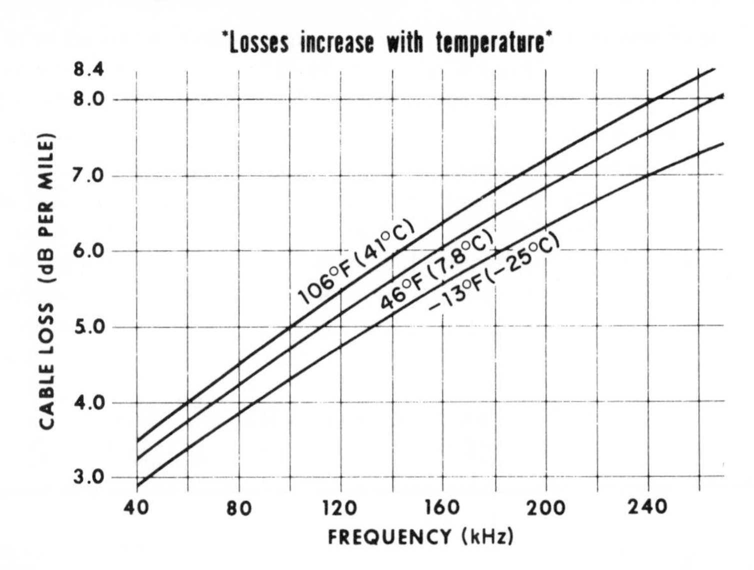

The design and operation of carrier systems are greatly influenced by cable characteristics. The prime factor is the increase in attenuation with frequency, known as *slope,* and the nonlinearity of this increase. Expressed in dB, slope also may be defined as the difference in signal level between two reference frequencies. For example, consider the GTE Lenkurt 47A/N1 carrier applied to a No. 19 gage toll cable pair. In the 36-140-kHz range (the low group of the N1 carrier), this cable pair has a loss of 4.5 dB per mile at 36 kHz, and about 7.5 dB per mile at 140 kHz. The slope is the *difference* between these two losses, or 3 dB per mile. For the usual repeater spacing of 8 miles, the loss will be 24 dB. The term *natural slope* refers to the method of equalizing a carrier system to compensate for a cable's *actual slope.*

The nonuniformity of cable attenuation with frequency along a slope line is termed *bulge.* It can be variously caused, for example, by irregularities in cable manufacture, improper splices, changes in cable gage, and faulty impedance matches. Since slope and bulge are even more pronounced traits in cable than in open-wire transmission lines, cable systems must be designed to compensate for both traits.

The considerable effect of temperature on losses in cable pairs depends on both climate and installation type. Underground cables experience rather small temperature variations, mostly seasonal. Aerial cables, however, being exposed, experience large variations, both seasonal and daily, sometimes from below freezing at night to over 90° F (32° C) the next day. The chart shows typical variations of attenuation with temperature for carrier frequencies normally used on cables. Although temperature changes also change the slope, more noticeably at frequencies below 40 kHz, their effect is not usually detrimental in short-haul systems operating above 40 kHz.

Slope Compensation Methods

Compensation for *slope* (the increase in attenuation with frequency) may be accomplished in cable carrier systems by three wholly different methods. First, pilot tones can be transmitted over the channel carrier to automatically adjust the gain-frequency characteristics of the receiving carrier channel. This had been employed on open-wire carrier systems. Another method is the use of fixed equalizers in the channel carrier equipment. These equalizers introduce a predetermined amount of attenuation that is caused to decrease with frequency across the carrier transmission band. This is termed *negative slope.* A third method, utilized in many cable carrier systems, is called *frequency frogging*—the inversion of the carrier frequency bands by modulation at each carrier repeater point.

The basis of the frequency frogging technique is that the highest frequency channel is shifted and inverted to the lowest carrier frequency channel before amplification. At the same time, the lowest frequency channel will be shifted and inverted to become the highest frequency channel. The same procedure is followed at the next frogging repeater, thereby restoring the channel frequencies and signal levels to the initial condition. The inversion of the carrier channels in successive repeater sections also minimizes the need for equalization. Since the higher frequencies have greater losses, the slope introduced

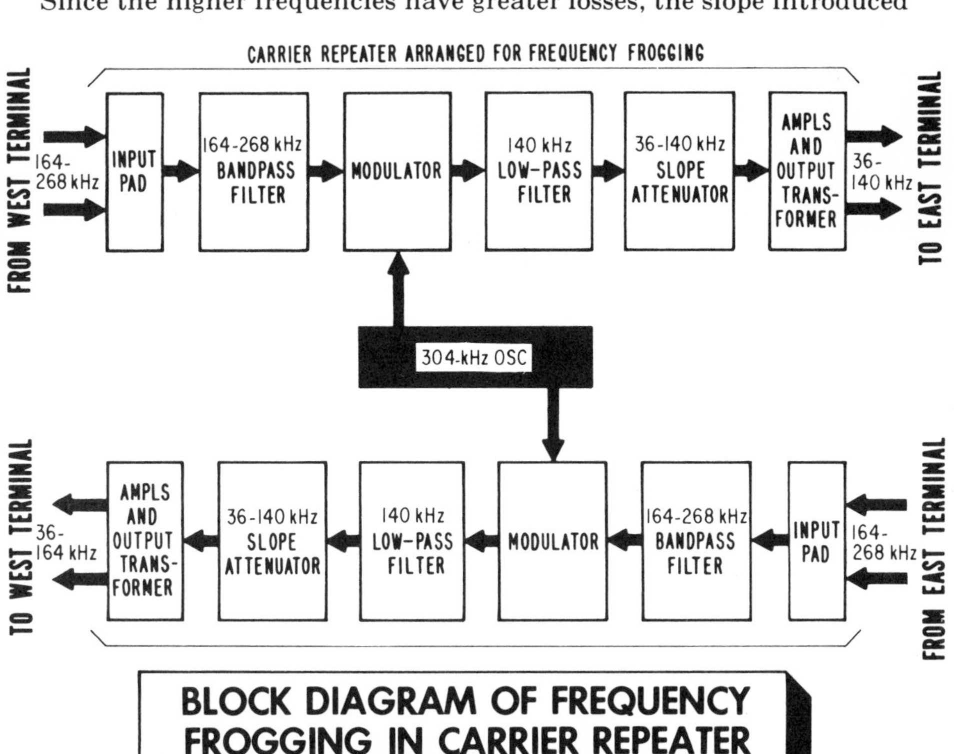

BLOCK DIAGRAM OF FREQUENCY FROGGING IN CARRIER REPEATER

Slope Compensation Methods (cont.)

by one repeater section will be effectively canceled by the reversed slope in the following section.

To understand this method better, let us examine the frequency allocations in a typical twelve-channel carrier system operating over No. 19 gage quad or toll cable. We will assume that channels Nos. 1 to 12 at the originating west terminal occupy the 164- to 268 kHz frequency region. From the chart on page 132, we are able to calculate the losses in the initial 8-mile repeater section. Channel No. 1, for example, will incur a loss of about 48 dB, whereas channel No. 12 will be attenuated 64 db. The losses in channels Nos. 2 through 11 will range more or less uniformly between these two values. The 16-dB difference in losses between channels Nos. 1 and 12 (64 dB – 48dB) is the slope at the first frogging repeater.

Let us now trace the progress of channels Nos. 1 and 12 through the frogging repeater as illustrated on page 135. Channels Nos. 1 to 12 from the west terminal enter the input pad of the frogging repeater and are passed through the 164- to 268-kHz bandpass filter to the modulator. These incoming channels are next modulated with the 304-kHz oscillator signal. Only the lower sideband products of the modulation process are permitted to go through the 140-kHz low-pass filter. Therefore, channel No. 1, which entered the repeater as 164 kHz, the lowest frequency, will leave it as the highest frequency (140 kHz) of the twelve channels. Similarly, channel No. 12, which came in as 268 kHz (the highest frequency), will leave as the lowest frequency channel (36 kHz). This will also take place for the twelve channels in the opposite (east-to-west) direction of transmission as shown.

You will recall that in the cable section to the first frogging repeater, channel No. 12 had the greatest loss because it was assigned the highest frequency. In the next cable section, owing to frequency frogging, channel No. 1 will have the highest frequency and, therefore, will suffer the most loss. Referring again to the chart on page 132, our calculations will show 44 dB loss for channel No. 1, (140 kHz) but only 28 dB loss for Channel No. 12 (36 kHz) in the second 8-mile repeater section. This 16 dB slope at the input to the second frogging repeater is reversed for channels Nos. 1 and 12 as compared to the first section. At the output of the second frogging repeater, channel No. 1 will shift to the lowest frequency (164 kHz) and channel No. 12 to the highest frequency (268 kHz). Thus all twelve channels will again be in the original frequency allocation order and at equal amplitude levels.

Another advantage of frequency frogging at repeaters is that the two transmission directions use different carrier frequency bands. The diagram shows that the 36- to 140-kHz and 164- to 268-kHz bands parallel each other for different transmission directions. This frequency separation prevents feedback (singing) and crosstalk interference at repeaters or other points where large level differences may exist.

Reducing Noise in Cables

The metallic sheath of a cable generally shields carrier circuits from external noise. Cables are subject, nevertheless, to various noises including carrier frequency noise. Some of these noise sources are unsoldered or poorly soldered cable splices, interchannel modulation and crosstalk from carrier systems, and dialing, signaling and other central-office-generated noise voltages. Cable splices made for normal voice frequency usage may not be soldered. Unsoldered splices may provide good connections for voice frequency currents but often cause noise in carrier circuits. Since copper oxide may be formed at unsoldered or poorly soldered joints of *dry* circuits (circuits in which direct current normally does not flow along cable conductors) and since it is a semiconductor, its build-up at splices can cause rectification of carrier currents. This condition can generate noise and give rise to other interference in the system. To overcome it, a small amount of direct current (generally termed *sealing current*) is applied to carrier cable pairs. This current prevents the formation of copper oxide at unsoldered splices or other poor joints. Carrier cable pairs normally carry direct current in order to power remote repeaters between carrier terminals as well as to serve as sealing current.

Carrier systems usually generate a certain amount of noise. Some may come from interchannel modulation and signaling tones. Similarly, considerable noise may enter the carrier circuit from the central office switching and signaling equipment over noncarrier pairs in the cable. To overcome this noise, it is often practical to install carrier repeaters near the noisy central office building (see diagram). This raises the received carrier levels high enough to override the noise. For short- and medium-haul carrier circuits, a *compandor* may be incorporated in the carrier channels. This device, described next, can substantially reduce the interfering effects of the noise.

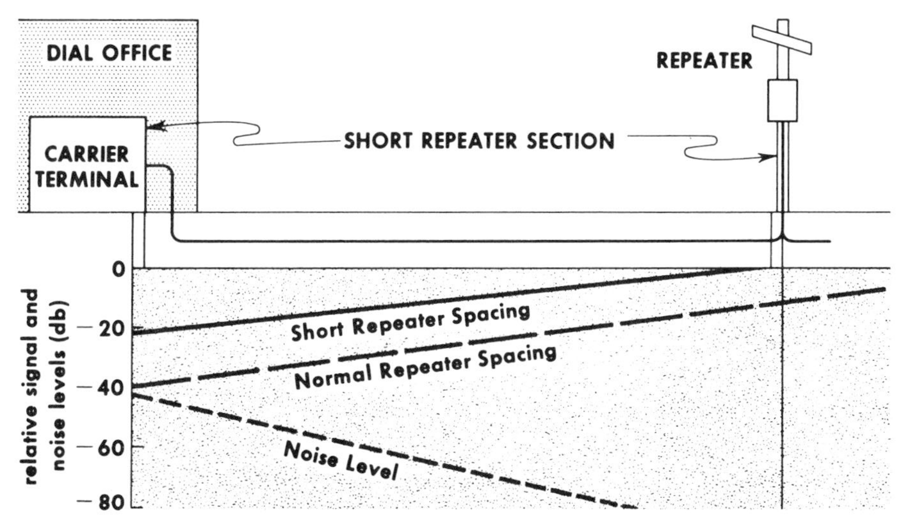

The Compandor

Noise introduced in the carrier cable pair between terminals can be substantially reduced without increasing the net loss of the circuit providing the transmitting carrier level is increased a specified amount and a loss of the same amount is inserted at the receiving end. This will cause the noise to be reduced by the same amount as the inserted loss. The device that can accomplish this signal-to-noise improvement is called the *compandor.* Its principle of operation is based on the volume range characteristics of speech. For instance, there may be as much as a 50-dB difference in sound power between the syllables spoken by a loud talker and a weak talker. The signals of the weak talker must be amplified for transmission at a higher level than the noise and crosstalk encountered. The amplification of the strong signals of the loud talker must be kept within limits to prevent overloading of amplifiers. The compandor permits transmission of speech intensities within a reduced range of levels. This considerably improves the signal-to-noise ratio of transmission.

The drawing shows the effect of the compandor in compressing and expanding speech signals. The compandor consists of two basic units—the *compressor* in the speech input circuit, and the *expandor* in the speech output section. In the drawing, it is assumed that the difference between the loud and weak talkers covers a range of about 50 dB. The compression action of the compandor in the input circuit can squeeze this range in half, or to 25 dB, for transmission. At the receiving end of the circuit, these signals pass through the expandor part of the compandor. The variable-loss device in the expandor reverses the previous action of the compressor in the transmitting compandor. The weakest speech signals receive maximum attenuation and the strongest signals are amplified slightly. Thus, the loss inserted by the expandor reduces the noise by the same amount. The signal-to-noise improvement effected in practice is usually about 21 dB.

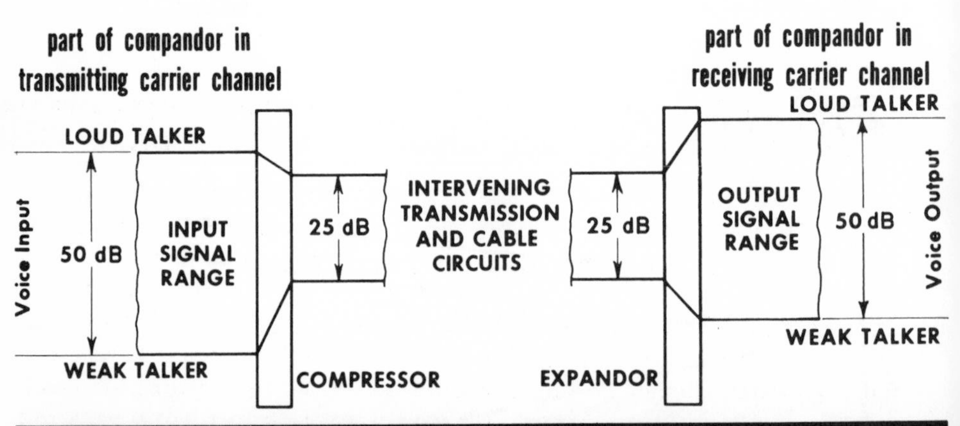

VOICE COMPRESSION AND EXPANSION ACTION OF COMPANDOR

Example of Compandor Performance

The compandor is employed to make noisy circuits satisfactory for toll transmission service. It must be understood, however, that the compandor is effective only if the noise is generated within the carrier equipment or the transmission medium, such as the cable pair. The compandor cannot reduce noise which is received along with the speech signals at the input to the carrier channel. Another important advantage of the compandor is the reduction of maintenance time needed for level adjustments. The compandor can provide a 21 dB reduction in the noise margin so that minor adjustments are not often required. This also increases the interval of making routine tests and measurements on carrier circuits. The diagram below illustrates the noise improvement effected by the compandor in a carrier channel.

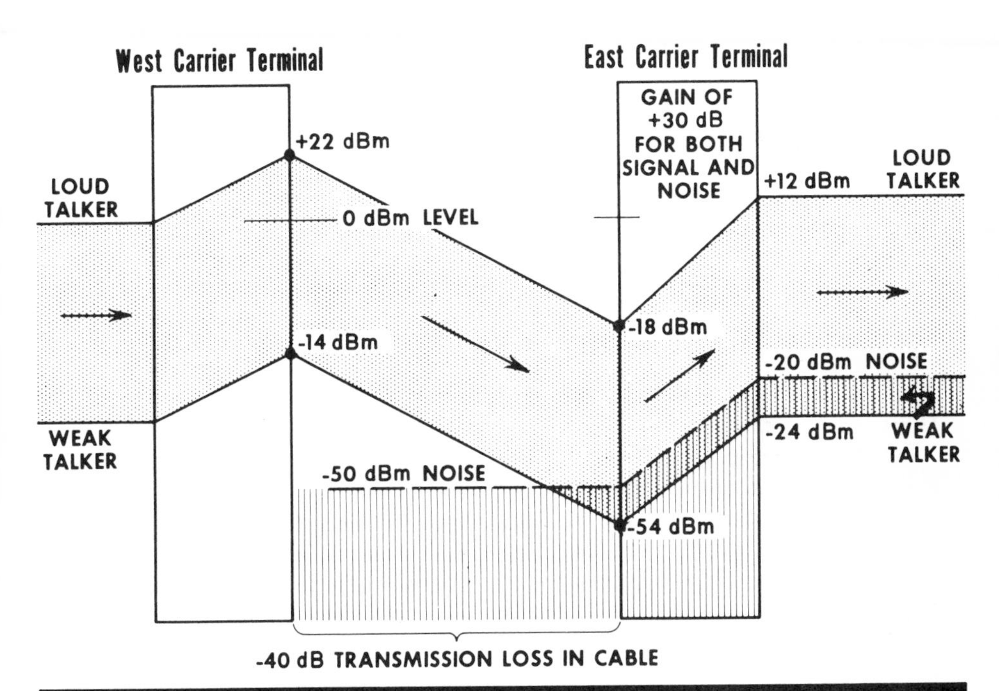

POWER LEVEL DIAGRAM WITHOUT COMPANDORS

The illustration is a power-level diagram of the signals on a carrier channel. This particular carrier channel is assumed to have a noise intensity of about −50 dBm (50 dB below 1 milliwatt) between two carrier terminals. The cable transmission loss between the terminals is 40 dB. Therefore, the signals of a weak talker would be received at a −54-dBm level. The noise would be 4 dB higher than this level. Without compandors, this circuit would not be satisfactory as a toll trunk.

Example of Compandor Performance (cont.)

Let us now install compandors, one at each end of the circuit before the channel unit.

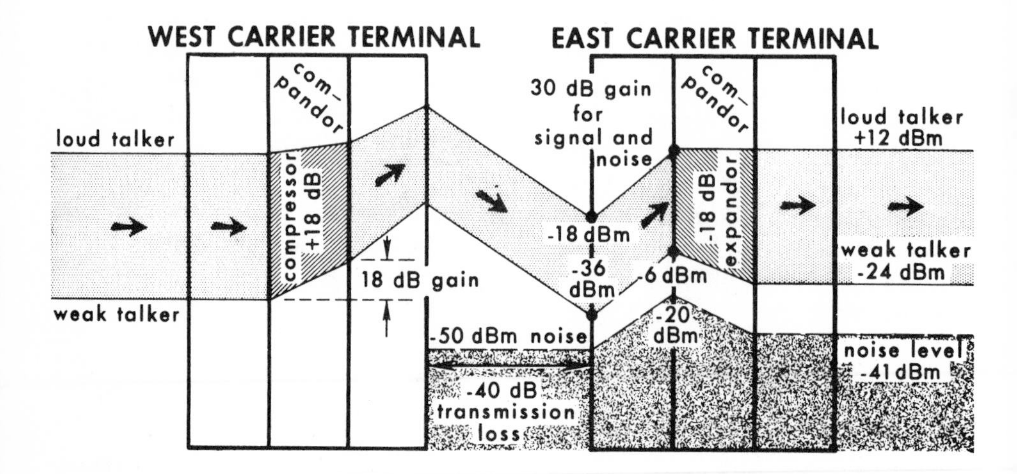

The power-level diagram is of the same carrier channel, now equipped with compandors. The speech signals will now first go through the compressor circuit of the compandor. The weak talker's signal is amplified about 18 dB in the compressor. The loud talker's signal is not amplified by the compressor. The signals reaching the East receiving carrier terminal are at levels of −18 dBm for the loud talker and −36 dBm for the weak talker. The line or cable noise is still −50 dBm as before. The carrier terminal equipment amplifies all signals and noise about 30 dB. They next pass through the expandor circuit of the compandor. The higher volume signal of the loud talker has a level of 12 dBm. It controls the expandor circuit. Therefore, the expandor introduces an attenuation of 18 dB to the weak talker's signal level of −6 dBm. This reduces its level to −24 dBm (24 dB below 1 milliwatt). At the same time, the noise is attenuated approximately 21 dB by the compandor action between syllables and words. This reduces the noise to −41 dBm (41 dB below 1 milliwatt). The net result is that the weak talker's signal, which was 4 dB below the noise level without a compandor, is raised 17 dB above the noise level. This shows the improvement possible in the voice signal-to-noise ratio that can be performed by the compandor.

It may be well, at this point, to state that compandors cannot be used on data transmission circuits. Data transmission comprises trains of binary digits (bits) in the order of microseconds or less. Impulse noise interference often has similar characteristics. As a result, the compandor may attenuate the data pulses in the same manner as for noise impulses.

Basic Terminal Equipment of Cable Carrier Systems

Modern FDM carrier systems, used with multiconductor cables, normally furnish up to 12 or 24 channels for toll-connecting and intertoll trunks. These carrier systems operate on a four-wire basis. Separate cable pairs and different frequency bands are utilized for each direction of transmission. As was explained in the section on Loading Limitations, nonloaded cables are used for these carrier systems.

Some representative 12-channel FDM cable carrier systems include: the Western Electric N1, N2, the GTE Lenkurt 47A/1, and 47A/2. These compatible carrier systems are of the double sideband, amplitude modulation, and carrier transmitted types. The 12 channels are spaced at 8-kHz intervals, in the low-group line frequency band of 36 to 140 kHz, for transmission over one cable pair. The high-group frequency band of 164 to 268 kHz is employed on another cable pair for the opposite transmission direction.

The Western Electric N3, and the GTE Lenkurt 45BN and 46B carrier systems, which are compatible, make use of the single sideband, suppressed carrier technique explained in the chapter, "Carrier Transmission and Signaling Features". Pilot frequencies, however, are transmitted for regulating line signal levels, carrier frequency synchronization, and related control functions. These carrier systems provide up to 24 trunk circuits, employing two cable pairs, one for each transmission direction. All aforementioned cable carrier systems can function over about 200 miles (125 km) of No. 19 gage nonloaded toll cable, utilizing frequency frogging repeaters spaced at approximately 8-mile (12.9 km) intervals.

The initial cable carrier systems, such as the GTE Lenkurt 45BN that utilized single sideband carrier suppressed operations, employed three or more modulation steps. These modulation steps were needed to raise the carrier channels from the voice band (250-3,400 Hz) to the required 40-140-kHz and 164-264-kHz frequency bands for transmission over cable pairs. The received carrier signals were demodulated down to voice band in the reverse order, using the equivalent number of modulation steps. Moreover, it was necessary to divide the 24 channels into six pregroup carrier units, each comprising four channel units. The various modulation and demodulation operations and related functions are subsequently described in detail.

Improved filter design, better frequency synchronization, and other technological advances now enable cable carrier systems to accomplish their frequency translation functions in two or three modulation steps. For instance, the drawing illustrates the modulation plan used for the 24 channels in the GTE Lenkurt 46B carrier system. The 24 voice channels are divided into two channel groups, each of 12-channel units. The 12 channels in each group are modulated by their respective channel carriers, as indicated in the drawing. The resultant lower sideband products of modulation are in the 60-168-kHz band for

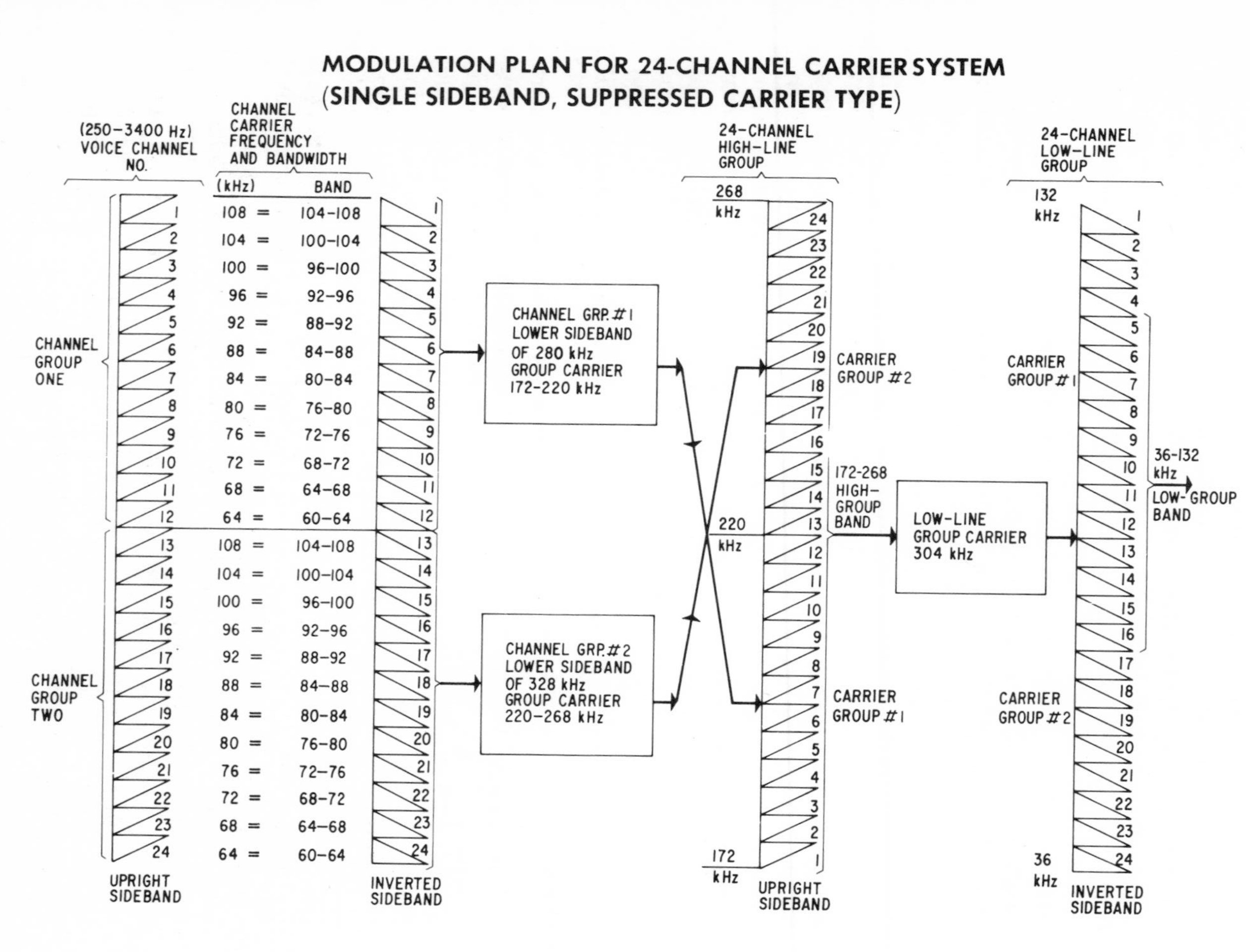
MODULATION PLAN FOR 24-CHANNEL CARRIER SYSTEM
(SINGLE SIDEBAND, SUPPRESSED CARRIER TYPE)
(250–3400 Hz) VOICE CHANNEL NO.
CHANNEL CARRIER FREQUENCY AND BANDWIDTH
(kHz) BAND
108 = 104–108
104 = 100–104
100 = 96–100
96 = 92–96
92 = 88–92
88 = 84–88
84 = 80–84
80 = 76–80
76 = 72–76
72 = 68–72
68 = 64–68
64 = 60–64
CHANNEL GROUP ONE
CHANNEL GROUP TWO
UPRIGHT SIDEBAND
INVERTED SIDEBAND
CHANNEL GRP. #1 LOWER SIDEBAND OF 280 kHz GROUP CARRIER 172–220 kHz
CHANNEL GRP. #2 LOWER SIDEBAND OF 328 kHz GROUP CARRIER 220–268 kHz
24-CHANNEL HIGH-LINE GROUP
268 kHz
220 kHz
172 kHz
CARRIER GROUP #2
CARRIER GROUP #1
172-268 HIGH-GROUP BAND
UPRIGHT SIDEBAND
LOW-LINE GROUP CARRIER 304 kHz
24-CHANNEL LOW-LINE GROUP
132 kHz
36 kHz
CARRIER GROUP #1
CARRIER GROUP #2
36-132 kHz
LOW-GROUP BAND
INVERTED SIDEBAND

Basic Terminal Equipment of Cable Carrier Systems (cont.)

both groups. Channel Group No. 1, consisting of channels numbered 1 to 12, is then raised by another modulation step (280-kHz group carrier) to the 172-220-kHz portion of the high-group frequency band. Likewise, channels 13 to 24 in Channel Group No. 2 are translated by the 328-kHz group carrier to the 220-268-kHz part of the high-group frequency range. The entire 172-268-kHz high-group band is next processed, as later explained, for transmission to a distant carrier terminal or to a repeater. To transmit the 36-132 kHz or low-group frequency band, the high-group frequencies (172-268 kHz) are further modulated by a 304-kHz group carrier, as shown in the drawing on the opposite page. This results in the formation of the 36-132-kHz low-group band. The various operations involved are similar to those described for the older type 45BN cable carrier system.

In order to present a simplified explanation of the several modulation and combining steps, filters, and amplification stages, and their functions in current FDM carrier systems; the operation of the older type GTE Lenkurt 45BN carrier will be described in detail. This should help the reader to better understand how present-day FDM cable carrier systems work. We will start with the diagram of the basic 24 channel units in the type 45BN carrier terminal.

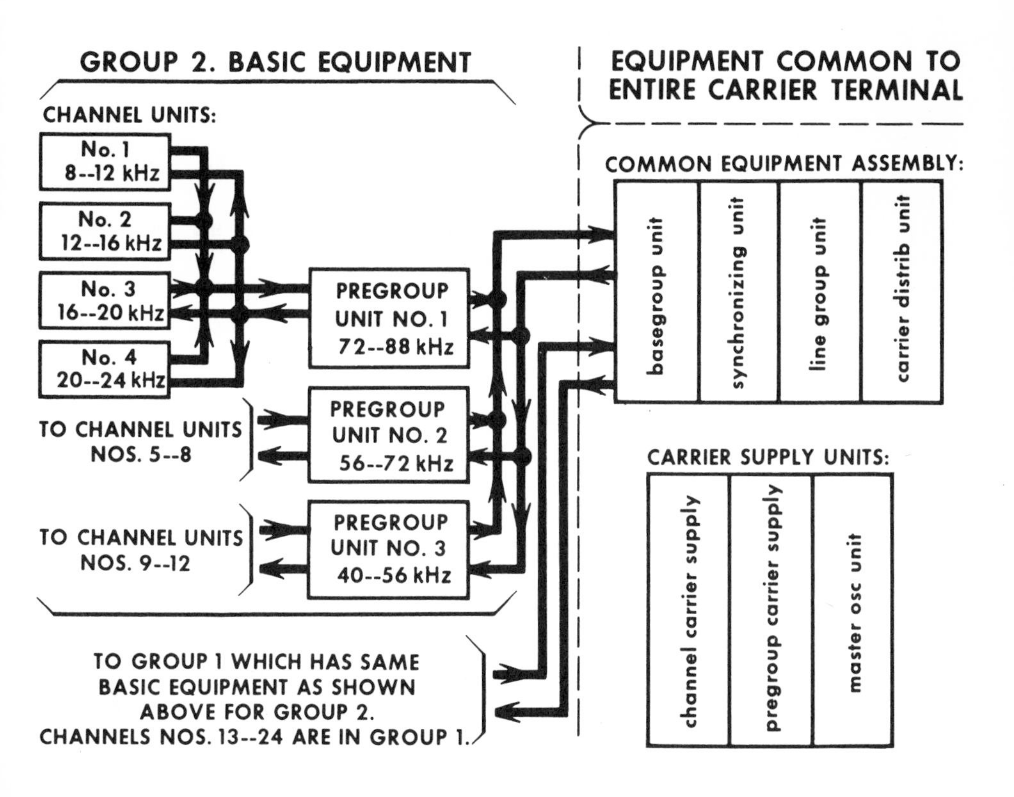

Cable Carrier Operations

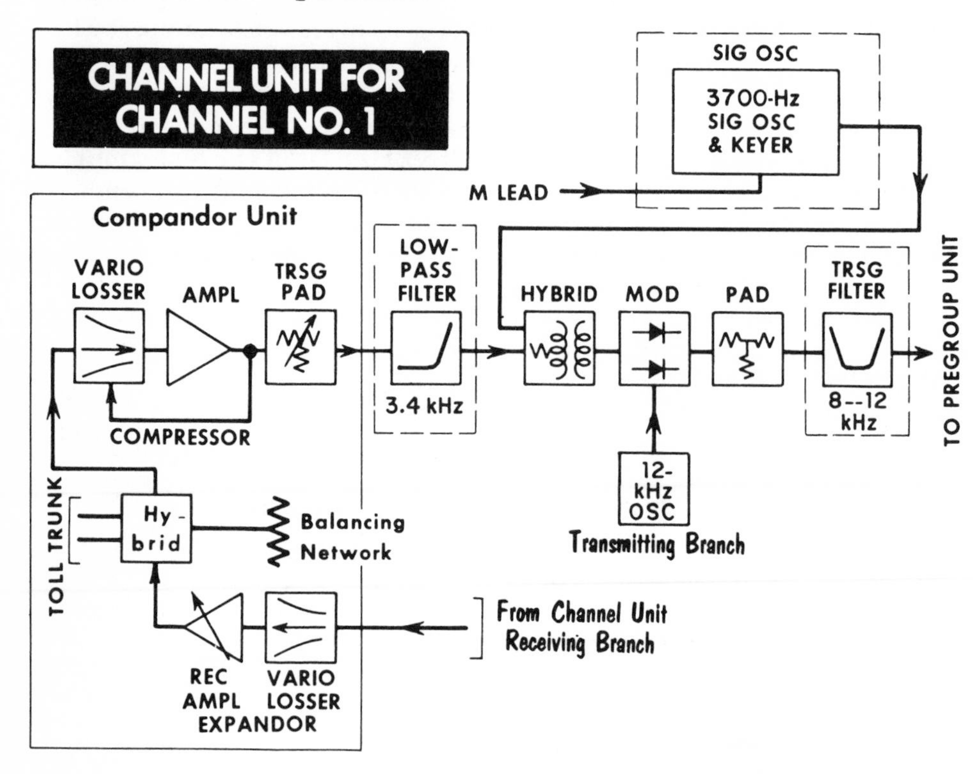

With the aforementioned basic equipment in mind, let us follow the progress of a speech signal through No. 1 channel of the old GTE Lenkurt 45BN cable carrier system.

Referring to the schematic diagram of a compandor and channel unit, note that the voice signals from the toll trunk first enter the compandor. The speech signals are compressed by the compandor action as discussed before. The voice frequencies then go through the low-pass filter of the channel unit. It restricts the voice signals to frequencies below about 3,400 Hz to allow for the use of the signaling circuit. The 3,700-Hz signaling frequency is combined with the speech signals in the adjacent resistance hybrid. The combined signals next enter the channel modulator. They are mixed with the 12-kHz carrier oscillator signals of the modulator. The output signals of the modulator are attenuated to the required level by the pad circuit. The resultant signals are now sent through the channel transmitting band-pass filter. This filter permits only the passage of the lower-sideband 8- 12-kHz signals. The other frequency products of modulation are greatly attenuated. The other three channel units, which make up a pregroup unit, operate in a similar fashion. Each channel unit, however, has a different carrier oscillator frequency.

Pregroup Unit

Each of the three *pregroup* units of a carrier group serves four channel units. The pregroup provides another modulation step for the channel carriers. It raises the frequencies of the carrier channels to the desired band for transmission. A pregroup unit, like the channel unit, consists of a transmitting and receiving branch. The transmitting branch comprises an 8-kHz high-pass filter, modulator, transmitting pad, and the transmitting bandpass filter. The receiving branch contains the receiving bandpass filter, demodulator, a 24-kHz low-pass filter and the receiving amplifier.

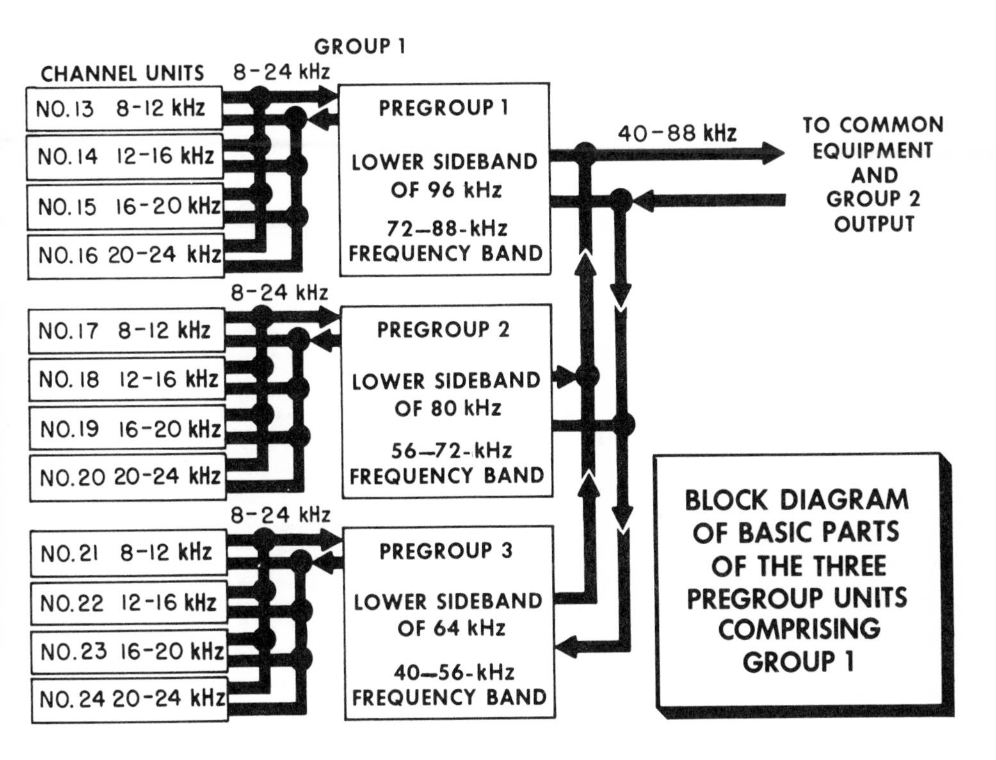

The diagram shows the connections and frequency relations between the 12 channel units and the 3 pregroup units that make up group 1 of the 45BN carrier system. These 12 channels are contained in the 40-88-kHz portion of the 40-140-kHz band to be sent over the cable pair. The other 12 channels of this 24-channel system are in group 2. The channel and pregroup units of group 2 are the same as in group 1. The three pregroup units of group 2, however, are further modulated by a 180-kHz carrier frequency in the *basegroup* unit as will be described later. This will place the 12 channels of group 2 in the 92-140-kHz band. The 12 channels of group 1 occupy the 40-88-kHz spectrum. Therefore, the entire 24 channels are in the 40-140-kHz range for transmission in one direction.

Transmitting Operations of Pregroup Unit

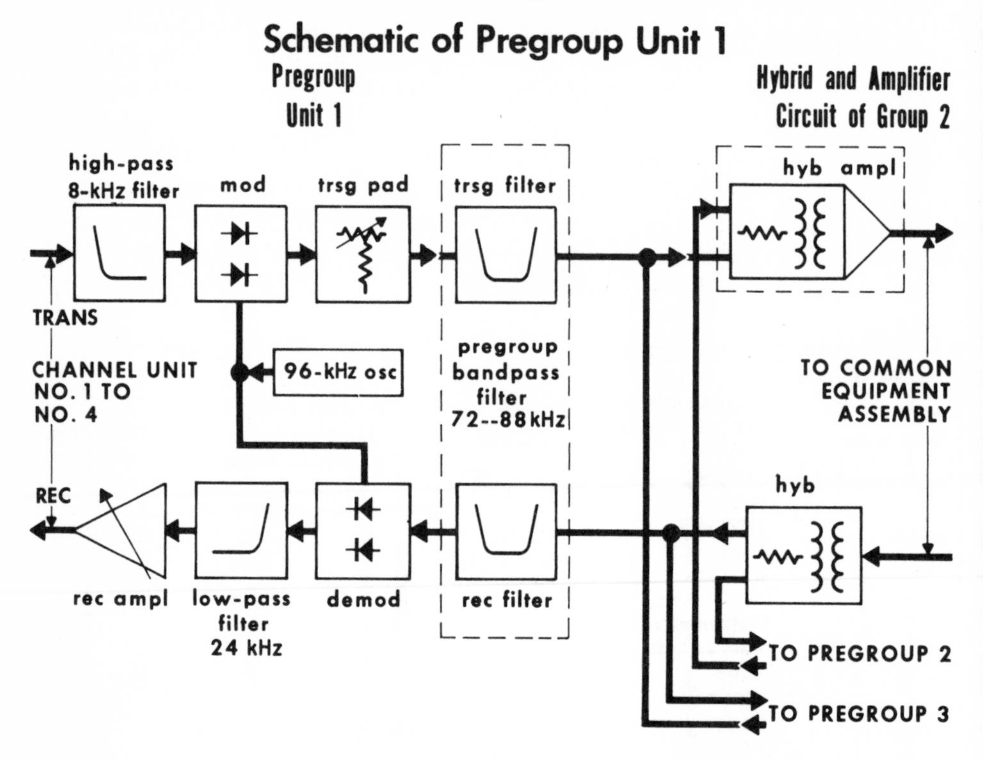

To continue with our carrier descriptions, the 8-12-kHz signals from channel 1 will enter pregroup unit 1. Referring to the schematic diagram, the signals go through the 8-kHz high-pass filter. This filter attenuates all frequencies below 8 kHz. The signals then are combined in the modulator with the 96-kHz carrier frequency. The pad attenuates the output of the modulator to the proper level before entering the 72-88-kHz transmitting bandpass filter. This filter permits only the lower sidebands of the 96-kHz modulation process to pass through. Therefore, the 72-88-kHz band which now contains the original 8-12-kHz frequencies of channel 1, proceeds unhindered to the hybrid and amplifier circuit of group 2.

These transmitting operations of the pregroup units may be condensed or restated as follows: The combined signals from channels 1 to 4 modulate a 96-kHz carrier in pregroup 1. The lower sideband of this modulation process is selected by the transmitting filter of this pregroup. The resultant 72-88-kHz band is passed to the group 2 amplifier circuit. A similar process is used in pregroup 2 where signals of channels 5 to 8 modulate an 80-kHz carrier. In pregroup 3, the signals of channels 9 to 12 modulate a 64-kHz carrier. These 12 channels, therefore, are positioned in the basic basegroup frequency range of 40 to 88 kHz.

Basegroup Operations

The *basegroup* unit is part of the common equipment assembly. It is used primarily for the 12 channels in group 2. These channels are numbered 1 to 12, inclusive. The 12 channels in group 1, which are designated numbers 13 to 24, inclusive, are connected directly through the basegroup unit to the synchronizing unit. The basegroup unit consists of a 180-kHz modulator and a 92-140-kHz bandpass filter in its transmitting branch. The receiving part includes the same type filter and a modulator connected to the same 180-kHz source.

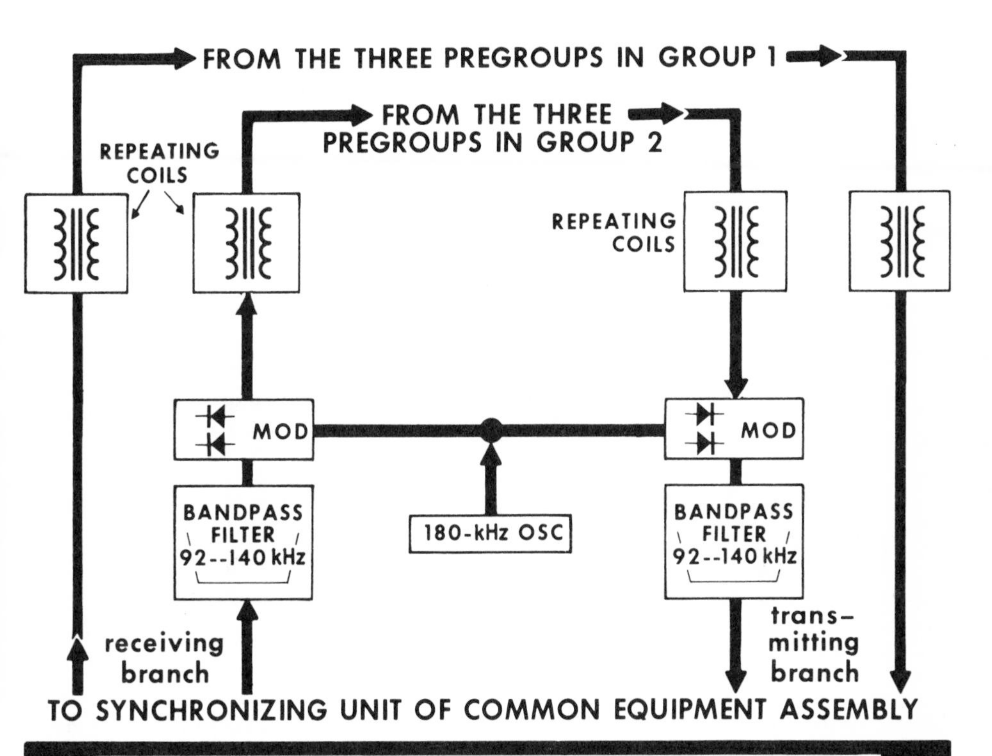

BASEGROUP UNIT OF COMMON EQUIPMENT ASSEMBLY

The signals under discussion originated in channel 1. They are now in the 72-88-kHz portion of the basic frequency range of 40-88 kHz. This band is further modulated by the 180-kHz modulator in the basegroup unit. The resultant sidebands are filtered by the transmitting bandpass filter. Only the lower sidebands can pass through this 92-140-kHz bandpass filter. Therefore, channels 1 to 12, comprising group 2, are now positioned in the 92-140-kHz portion of the frequency spectrum. Channels 13 to 24, comprising group 1, occupy the 40-88-kHz band as previously discussed.

Transmitting Functions of Synchronizing Unit

The next stop for our signals is the *synchronizing* unit of the common equipment assembly. The transmitting portion of the synchronizing unit includes attenuation pads and hybrid network for combining the two 12-channel carrier groups, group 1 and group 2; amplifier; equalizer to compensate for about 8 dB of cable slope; and a 96-kHz synchronizing pilot carrier introduced through a pad circuit. The receiving branch contains a considerable amount of equipment, which is described later.

DIAGRAM OF TRANSMITTING BRANCH OF SYNCHRONIZING UNIT

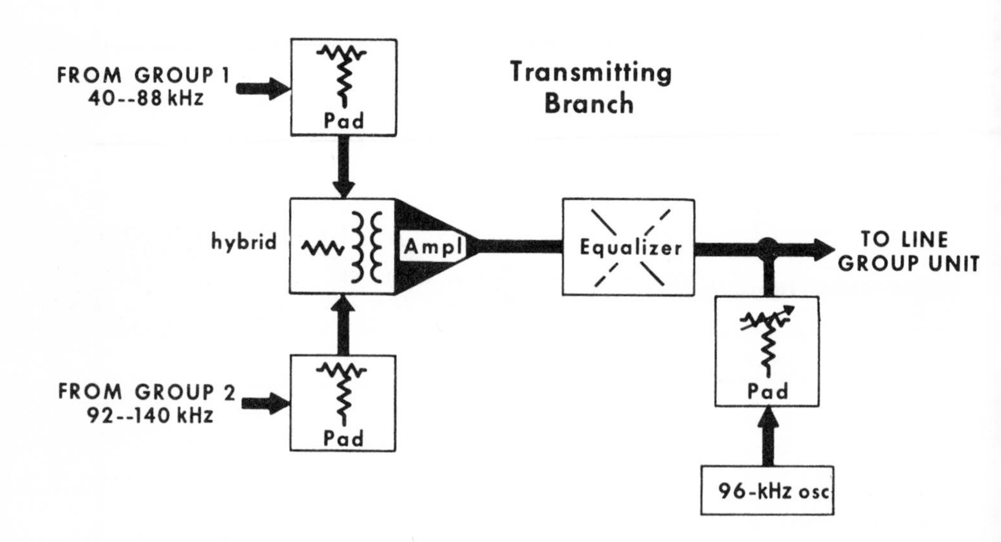

The signals in our example are now in the 92-140-kHz range as part of group 2. They enter the synchronizing unit and are combined therein with the 12 channels (13 to 24) of group 1. These latter signals are in the 40-88-kHz range. The result will be a 24-channel carrier group occupying the 40-140-kHz spectrum for one direction of transmission. This is called the low group because it contains the lowest carrier frequencies of the system. The entire carrier group is amplified and pre-equalized. The equalizer circuit is designed to provide compensation for about 8 dB of cable slope. A 96-kHz synchronizing pilot signal is next introduced into the transmission path. This 96-kHz pilot tone is used for both synchronization and amplifier regulation. It provides synchronization control to the distant carrier terminal, and also is used to regulate the repeaters and the distant terminal amplifiers for variations in the cable attenuation.

Transmitting Functions of Line Group Unit

The *line-group unit* provides two alternate arrangements. If the low group of 40-140-kHz frequencies is to be transmitted over the cable pair, the signals are routed through the 40-140-kHz bandpass filter and pad circuit. This circuit suppresses the undesired out-of-band frequencies. The 40-140-kHz signals are subsequently amplified by the transmitting amplifier and sent over the cable pair to the distant terminal.

The diagram is a schematic of a line group unit showing the two transmitting arrangements. If it should be desired to transmit the high group instead of the low group frequency band, use is made of the 304-kHz modulator. In this case, the signals first pass through a 140-kHz low-pass filter before being modulated by the 304-kHz carrier signal. The 164-264-kHz bandpass filter, following the modulator, selects only the lower sideband. The other frequencies are attenuated. This 164-264-kHz frequency band is then amplified and sent along to the cable pair for transmission to the distant end. The carrier distributing unit is adjacent to the line group unit. It includes repeating coils with simplex connections for feeding direct current to power remote repeaters.

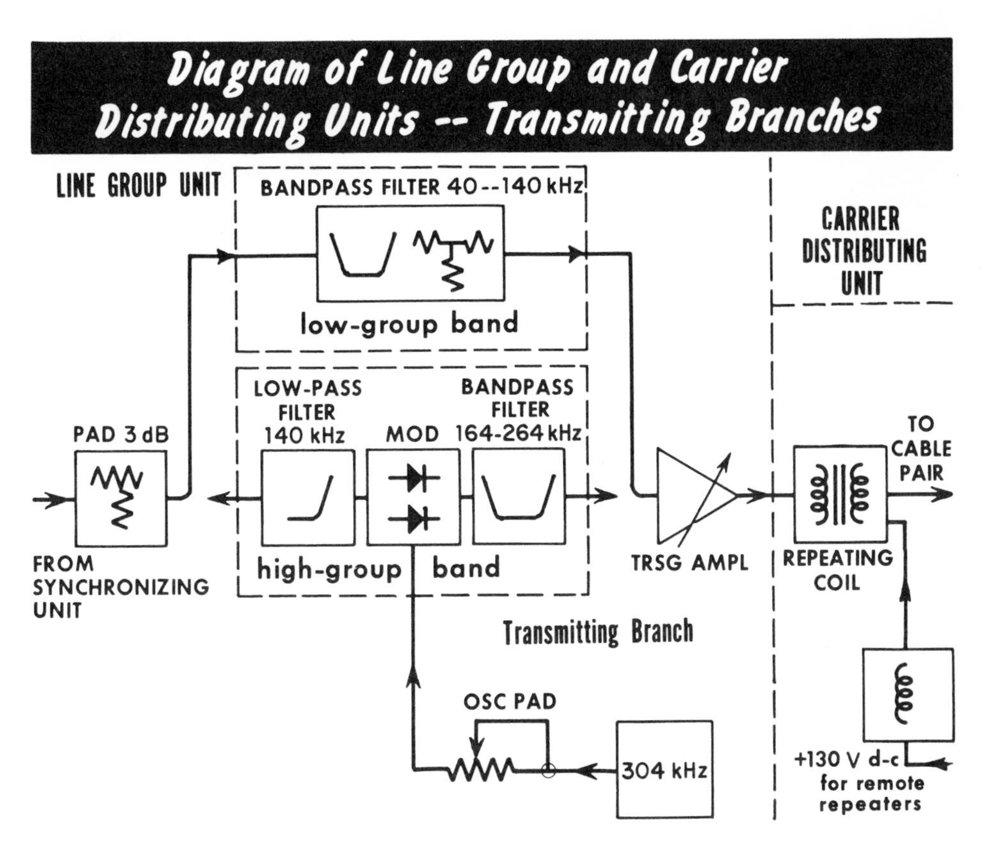

Receiving Functions of Line Group Unit

We will now discuss the reception of the signals at the distant carrier terminal. The signals first pass through the artificial line networks, repeating coil, and span pads in the carrier distribution unit. The span pad and artificial line network may be adjusted as required for the particular carrier installation. The received band of carrier frequencies is passed to the receiving part of the line group unit.

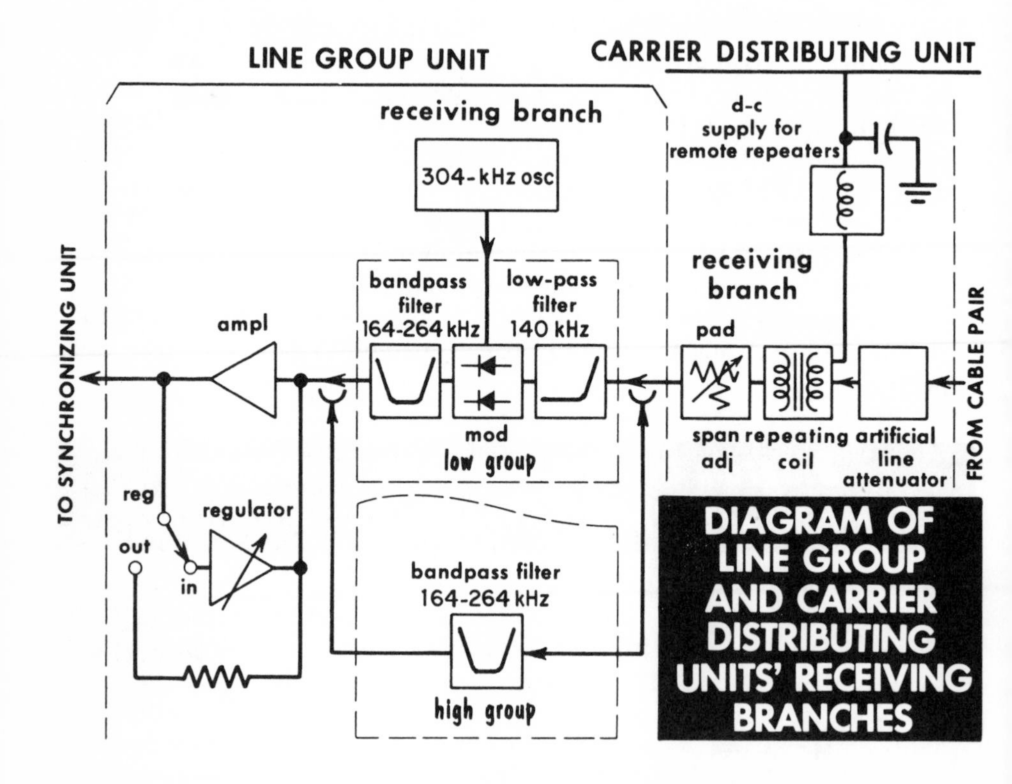

DIAGRAM OF LINE GROUP AND CARRIER DISTRIBUTING UNITS' RECEIVING BRANCHES

The low group, or 40-140-kHz frequency band, in which our channel 1 is located, goes through the 140-kHz low-pass filter. The frequencies above 140 kHz are eliminated by this filter. The signals next are modulated by the 304-kHz carrier frequency in the modulator circuit. The 146-264-kHz bandpass filter which follows permits only the lower sidebands of the modulation process to pass. These 164-264-kHz signals are amplified and regulated. The regulator circuit is under control of the 208-kHz signal which was originally the 96-kHz synchronizing pilot. The 96-kHz carrier signal had been changed to 208 kHz as a result of the previously mentioned modulation process. The alternate path through the 146-264-kHz bandpass filter is used for direct reception of the high group or 164-264-kHz band of frequencies. Subsequent operations are the same as for the low group frequency band.

Receiving Functions of Synchronizing Unit

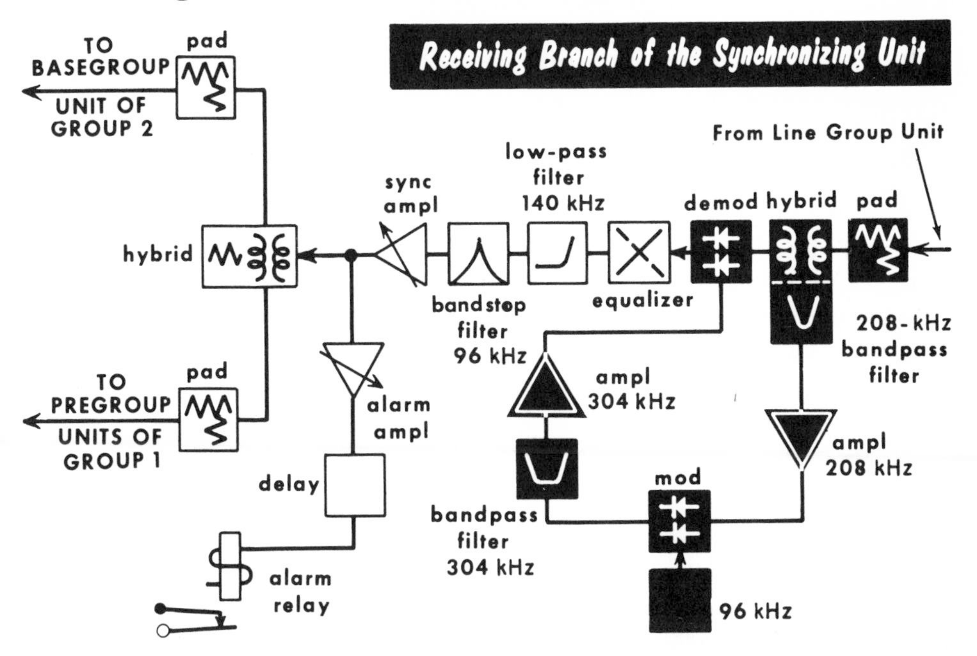

The receiving branch of the synchronizing unit consists of several different equipment items. They include hybrids, 208-kHz and 304-kHz bandpass filters, modulator, demodulator, 140-kHz low-pass filter, amplifiers, and attenuation pads. The schematic diagram illustrates the circuit arrangements.

The carrier signals from the line group unit proceed through the pad and hybrid network. At this point, the 208-kHz bandpass filter permits the 208-kHz synchronizing pilot signal to be extracted and amplified. It is then modulated by the 96-kHz local carrier oscillator. The 96-kHz carrier signal originated from the oven-controlled crystal oscillator of the master oscillator circuit. Therefore, its frequency is stable and exact. The resultant 304-kHz modulation product is filtered from the sidebands by the 304-kHz bandpass filter. After amplification, it is used to demodulate the 164-264-kHz band of signals from the line group unit. Let us now discuss the synchronizing functions performed in the synchronizing unit.

The 304-kHz signal, referred to above, actually was derived from the 96-kHz pilot signal originally transmitted from the distant carrier terminal. Therefore, any frequency shifts in the 96-kHz or 304-kHz oscillators in the distant carrier terminal or intervening repeaters will be reflected in this derived 304-kHz carrier or pilot signal. For example, assume that the frequency of the incoming synchronizing pilot signal was 95.9 kHz instead of 96.0 kHz. After combining with the 304-kHz local carrier frequency (crystal-controlled) in the line group unit, the derived lower-sideband frequency will be 208.1 kHz.

Receiving Functions of Synchronizing Unit (cont.)

This 208.1-kHz frequency will be mixed with the 96-kHz carrier frequency in the modulator of the synchronizing unit. The resulting product will be 304.1 kHz. The carrier signals from the line group unit will be in the 39.9-139.9-kHz spectrum because of the aforementioned frequency shift of the pilot signal to 95.9 kHz. After modulation with the 304-kHz carrier in the line group unit and filtering, these signals were changed to the 164.1-264.1-kHz range. Now, these signals will be demodulated in the synchronizing unit by the 304.1-kHz carrier signal. The net result will be a conversion to the 40.0-140.0-kHz band. In this manner, automatic synchronization of the demodulation frequencies will be obtained.

Let us now return to the schematic diagram of the synchronizing unit and follow the path of the 164-264-kHz signals as they pass through the hybrid to the demodulator. The 304-kHz signal demodulates them to 40-140 kHz for the lower sideband. The equalizer provides the necessary slope equalization for the effects of the cable pair. The 140-kHz low-pass filter eliminates all frequency products above 140 kHz. The following 96-kHz band-stop or band-elimination filter eliminates the 96-kHz synchronizing pilot signal from this 40-140-kHz band. The synchronizing unit amplifier amplifies this 40-140-kHz band before it enters the hybrid arrangement. This particular network is designed to divide the carrier signals into two equal parts. One part goes to the basegroup unit for further demodulation before entering the pregroup units of group 2. The other part will go directly to the pregroup units of group 1.

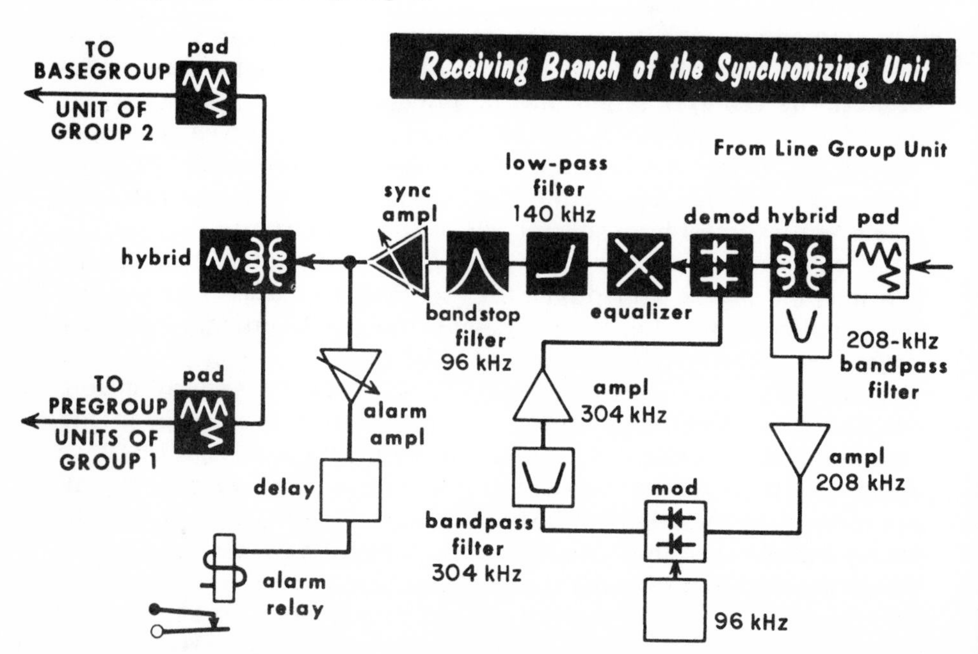

Receiving Functions of Basegroup and Pregroup Units

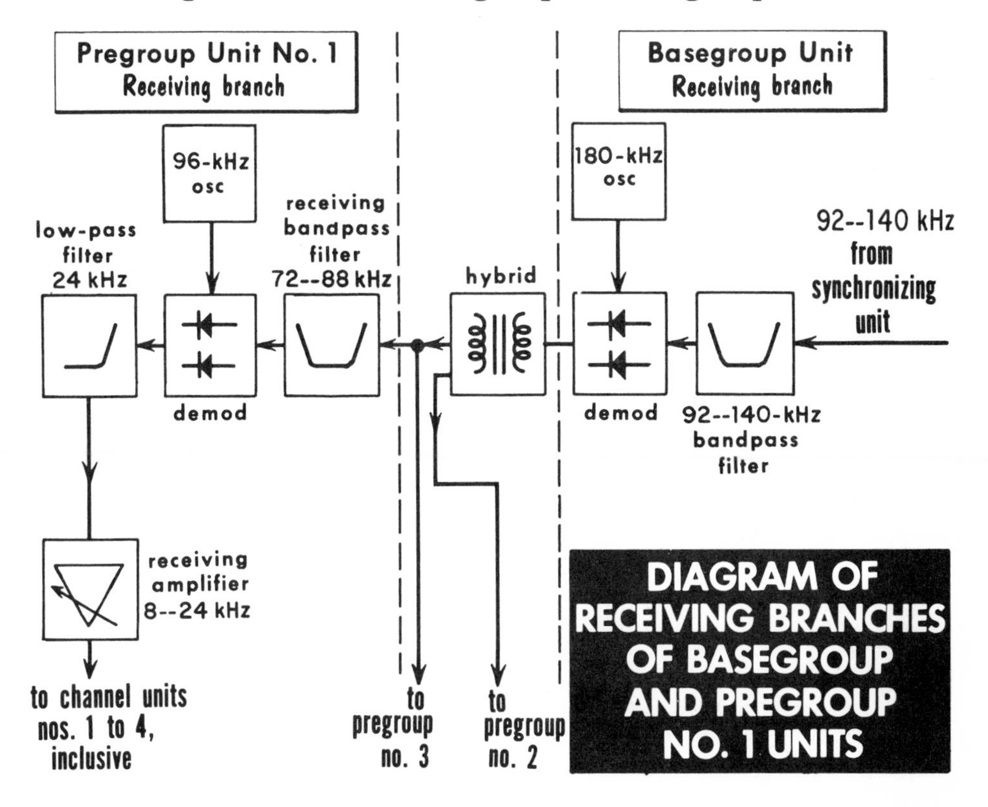

As previously stated, the carrier channel (channel 1) of our example is in the 12 channels comprising group 2. These channels occupy the 92-140-kHz, or low-group band of group 2. Therefore, the 92-140-kHz bandpass filter in the receiving branch of the basegroup unit selects this band of frequencies for further demodulation. The 180-kHz carrier oscillator and demodulator in the basegroup unit changes this 92-140-kHz band of 12 carrier channels to the 40-88-kHz range.

Referring to the schematic diagram, we see that this band of frequencies is impressed upon each of the three pregroup units of group 2. Each pregroup unit, as we previously learned, serves four channel units. The 72-88-kHz receiving bandpass filter of pregroup 1 permits the passage of this frequency band to its demodulator circuit, which is controlled by the 96-kHz oscillator. The 24-kHz low-pass filter after the demodulator effectively eliminates all frequencies above 24 kHz. After amplification by the receiving amplifier, the resultant signals are in the 8-24-kHz band. These carrier signals now are applied to the four parallel-connected channel units which make up pregroup 1. These are the channel units that are numbered 1 to 4, inclusive.

Receiving Functions of Channel Unit

The 8-12-kHz bandpass filter in the receiving branch of channel 1 selects this band of frequencies from the 8-24-kHz range emitted by pregroup 1.

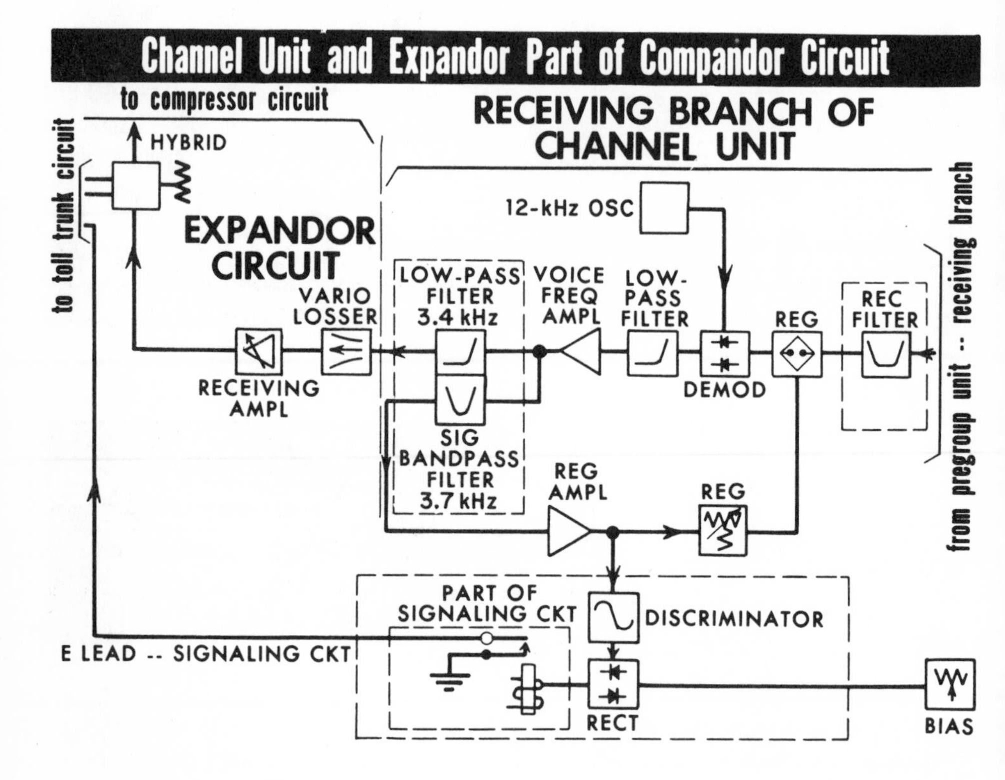

The signals next pass through the channel regulator to the 12-kHz carrier demodulator. The channel regulator is controlled by the 3,700-hertz signaling tone, as discussed in the section on Out-of-Band Signaling System. After demodulation, the low-pass filter suppresses the frequencies above about 4 kHz. The voice frequency amplifier following this low-pass filter raises the signals to the desired level. The signals now enter the 3.4-kHz low-pass and the 3.7-kHz bandpass filters. They separate the signaling tones from the speech signals which had been compressed in the distant-end compandor. They next go to the expandor part of the compandor which is associated with the channel unit. The variable-loss device in the expandor affords maximum attenuation to the weakest signals and a slight gain to the strongest signals. In this manner, whatever loss is inserted by the expandor reduces the noise by the same amount. (This action has been described.) The speech signals now complete their journey by entering the toll trunk circuit. The compandor action has resulted in an improvement in the signal-to-noise ratio of about 21 dB.

Carrier Supply and Master Oscillator Units

The carrier oscillator frequencies for the modulators and demodulators in the 24 channel units are supplied by the *channel carrier supply unit.* These frequencies are 12 kHz, 16 kHz, 20 kHz, and 24 kHz. The 96-kHz, 80-kHz and 64-kHz carrier oscillator frequencies for the modulators and demodulators in the pregroup units are furnished by the pregroup carrier supply unit. The primary source for all these carrier frequencies is the *master oscillator unit.* This contains two oven-controlled crystal oscillators which generate frequencies of 96 kHz and 180 kHz, respectively.

CARRIER SUPPLY UNITS AND MASTER OSCILLATOR OF LENKURT TYPE 45BN CABLE CARRIER SYSTEM

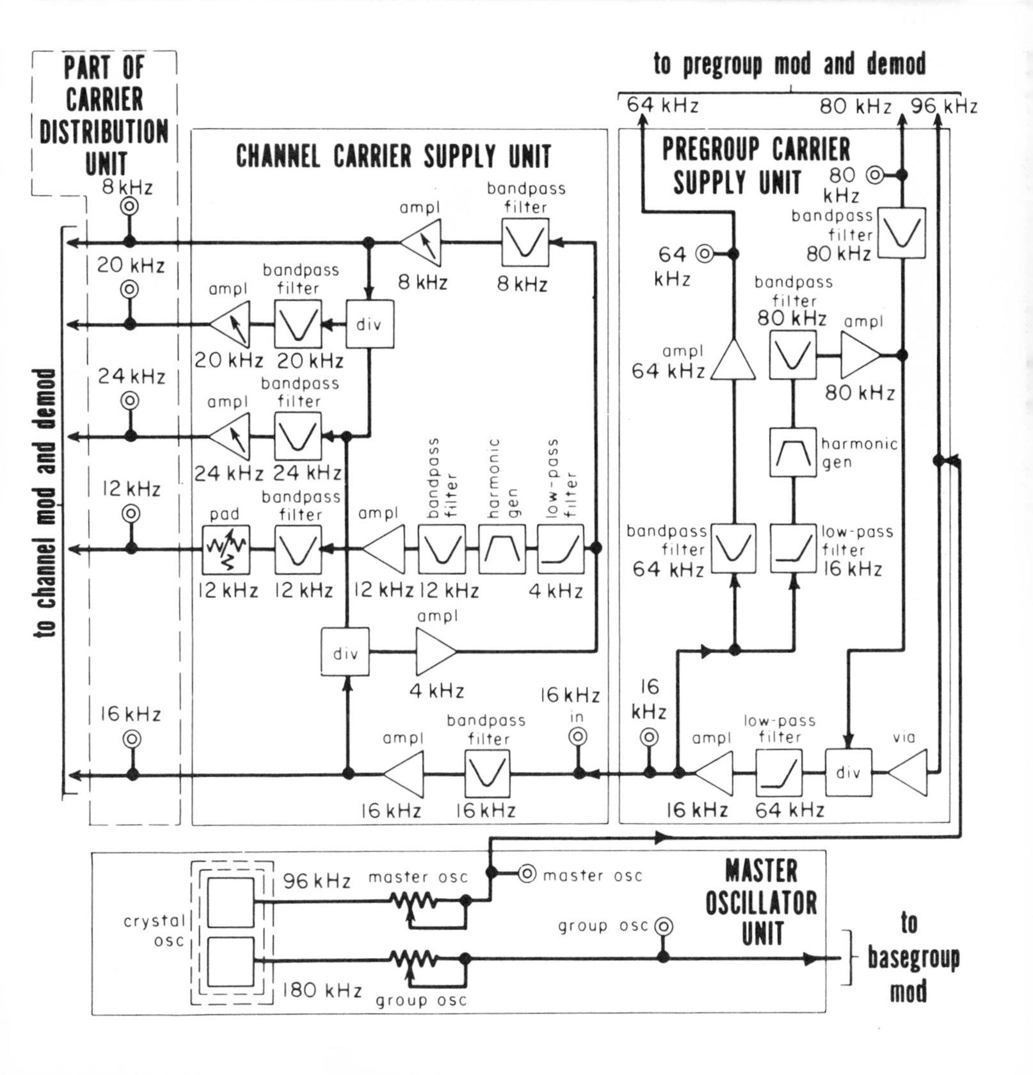

Carrier Supply and Master Oscillator Units (cont.)

The 96-kHz carrier frequency of the master oscillator unit provides the following functions: 96-kHz modulating and demodulating frequencies for pregroup 1; 96-kHz synchronizing pilot for the carrier system; regulation of amplifier gains in repeaters and at the far-end carrier terminal; a portion of this 96-kHz energy provides a controlled excitation voltage to the harmonic generation circuits in the pregroup and channel carrier supply units. This permits the generation of the two other pregroup frequencies of 80 kHz and 64 kHz in the pregroup supply unit. It also enables the channel carrier supply unit to generate the channel modulation and demodulation frequencies of 12 kHz, 16 kHz, 20 kHz, and 24 kHz. A separate 304-kHz crystal-controlled oscillator is provided in the line group unit to position the 24 channels in the 164-264-kHz transmitting band. This high-group band is used for transmission in the opposite direction to the 40-140-kHz frequency band.

Questions and Problems

1. What is the main cause for the attenuation of carrier frequencies in telephone cables?
2. What is the relationship between impedance and carrier frequency characteristics of a cable pair? Why?
3. Define slope. What causes bulge?
4. Explain the frequency frogging technique. What is its main advantage?
5. Is it necessary to solder all splices on cable pairs used for carrier circuits?
6. What is the purpose of a compandor? What is its principle of operation?
7. Name some representative cable carrier systems used in the toll network.
8. What approximate frequency bands are utilized by both the GTE Lenkurt 46B and the Western Electric N1-type carrier systems? Why is there a difference in channel capacity?
9. What controlled synchronization and amplifier regulation in the old Lenkurt 45BN carrier system?
10. What channel carrier oscillator frequencies were generated in the Lenkurt 45BN carrier terminal? What was the primary source of all carrier frequencies?

Limitations of Frequency Division Carrier Systems

The cable carriers that we have studied have utilized the Frequency Division Multiplex (FDM) technique. Filters of exact prescribed characteristics and highly linear amplifiers are indispensable elements of these carrier systems. If an amplifier is not exactly linear, considerable distortion and noise will be introduced into all carrier channels. For example, a strong signal in one channel of a 24-channel cable carrier may cause an amplifying stage to operate in its nonlinear region and degrade the transmission features of all 24 channels. In fact, the principal limitation to the number of channels in FDM carrier systems is not bandwidth but the loading or linearity characteristics of amplifiers.

The number of repeaters that are used also has a limiting effect on FDM carrier systems. Each repeater amplifies the noise, crosstalk, and other distortion products that may have accumulated in the previous cable section. Precise filters are also required in each repeater when the *frequency frogging* method is employed to compensate for the increased attenuation in the cable with frequency.

Improvements in the state of the art and the introduction of solid-state devices, such as transistors, substantially reduced the costs of carrier equipment. As a result, the use of carrier was extended to shorthaul toll circuits and even to subscriber lines in rural areas. Economic considerations initially precluded the application of FDM type carrier systems to interoffice trunk cables as a means of supplanting voice-frequency circuits. Technological improvements in the Time Division Multiplex (TDM) and Pulse Code Modulation (PCM) methods subsequently changed this situation, as we will perceive. The drawing shows the basic elements of FDM and TDM systems.

BASIC ELEMENTS

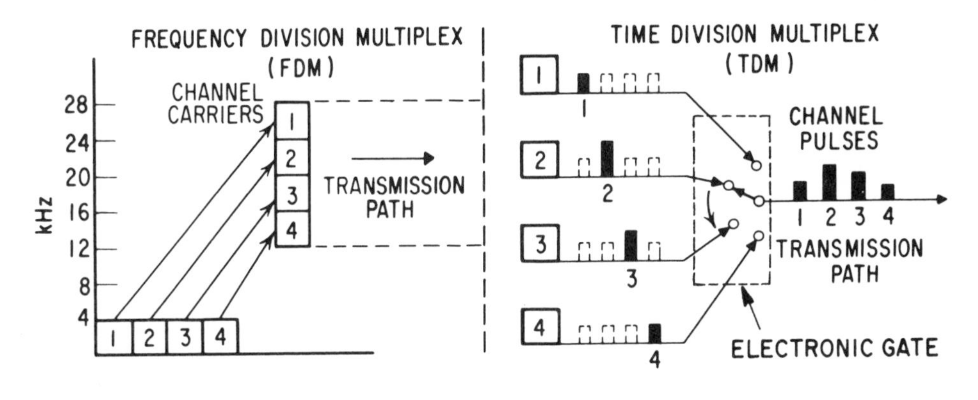

■-CHANNEL SAMPLING PULSES. □-SAMPLING INTERVALS OF OTHER CHANNELS.

Time Division Multiplex Fundamentals

The transmission of information, such as speech signals, by a carrier system can be represented as a voltage wave that varies continuously with time. This variation is the carrier modulating function in the FDM system. The carrier is varied continuously in an *analog* manner using amplitude (AM) or frequency modulation (FM). This continuous transmission of information describing the modulating function is not necessary in *Time Division Multiplex (TDM)* systems with pulse modulation. The drawing illustrates this relationship by depicting a voltage varying with time as in a speech wave. A series of pulses, each proportional to the amplitude of the continuous wave and corresponding in time to it, is shown beside the voltage wave.

SPEECH VOLTAGE WAVE AND EQUIVALENT PULSES REPRESENTING ITS AMPLITUDE

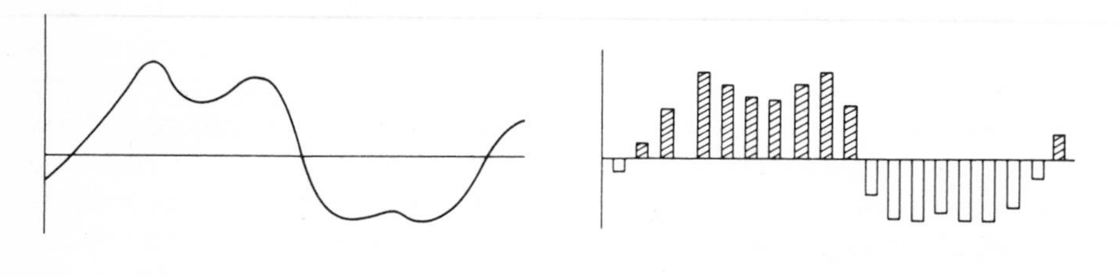

SPEECH VOLTAGE **PULSE AMPLITUDES**

A series of pulses may thus be used to carry information instead of a continuous modulated carrier. That is, the number of pulses transmitted per unit of time represents a sampling of the information in the voltage wave. Pulse duration, normally measured in microseconds, is determined by equipment and design limitations.

At this point, it is appropriate to inquire about the minimum number of pulses (or samples) required to represent the original signal faithfully. Let us refer to the sampling theorem of time division systems for the answer. Its basic principle essentially states that *if a message or signal is sampled instantaneously at regular intervals and at a rate which is at least twice the highest significant frequency of the signal, then the samples will contain all the information of the original signal.* Another version is that samples taken every t seconds will represent the signal as accurately as if it had been transmitted through a filter of bandwidth $t/2$. Thus, a particular voice signal can be represented adequately by taking samples at a rate equal to twice the highest signal frequency.

TDM Transmission Methods

Let us consider some representative TDM sampling examples. A single tone, such as a 1,000-Hz signal, will require sampling at the rate of 2,000 Hz. Voice-frequency (VF) signals in the 200-3,400-Hz band would necessitate a minimum sampling rate of 6,800 Hz. Since VF channels usually occupy a 4,000-Hz (4-kHz) bandwidth, an 8,000-Hz sampling rate is utilized to insure faithful reproduction of all voice currents. Samples of voice signals from many different VF channels can be transmitted along the same wire pair on a time-shared basis.

The 8,000-Hz sampling rate affords an interval of 125 microseconds ($t = 1/f$) between each sample of a particular voice signal. The sampling pulses, consequently, will be extremely short, in the order of a few microseconds, depending upon the number of voice channels to be sampled each 125 microseconds. If, for instance, 24 separate voice channels were transmitted over the same cable pair (each sampled at the 8,000-Hz rate), all 24 channels would be sampled at a combined rate of 192,000 times per second (8,000 Hz per channel $\times$ 24 channels). Each sample, therefore, would have its own separate time slot of 1/192,000 or a maximum duration of 5.2 microseconds. Successive TDM samples of different voice signals would go into progressive time slots. For example, the samples from the first voice channel would go in time slots No. 1, No. 8,001, No. 16,001, No. 24,001, No. 32,001, etc. Samples representing signals in the second channel would appear in time slots No. 2, No. 8,002, No. 16,002, No. 24,002, No. 32,002, etc. Similarly, in a 24-channel TDM carrier system, samples of the last channel would be in time slot No. 24, No. 8,024, No. 16,024, No. 24,024, No. 32,024, etc.

For a sampling rate of 8,000 Hz, the time between successive samples of any channel is 125 microseconds (1/8,000). This 125 microsecond interval is the *frame* period during which all channels must be sampled and processed as subsequently explained. The first channel is then sampled again and the same sequence is repeated. At the receiving terminal, the incoming TDM signals or sampling pulses must be switched in equivalent sequence. This switching action necessitates exact synchronism with that at the transmitting carrier terminal, thus ensuring that each of the 24-voice channels will receive its properly assigned speech samples.

The drawing on the following page shows in simplified form a block diagram of an elementary TDM carrier system with pulse amplitude modulation. The timing or sampling gate in the transmitting terminal connects to each of the 24 voice channels once every 125 microseconds. This time period corresponds to a sampling rate of 8,000 Hz. The resultant pulses of voice channel No. 12, which are illustrated in the drawing, are variations in the amplitude of the speech signal with time. These pulses are amplified and sent over the cable pair to the distant receiving terminal. The timing gate at the receiving end is

TDM Transmission Methods (cont.)

in exact step with its counterpart at the transmitting terminal. Therefore, it will be connected to voice channel No. 12 at the same time. The received pulse amplitude modulations will consequently be reconstituted into the original speech wave existing in voice channel No. 12. Speech transmissions in the reverse direction will follow a similar procedure on another cable pair.

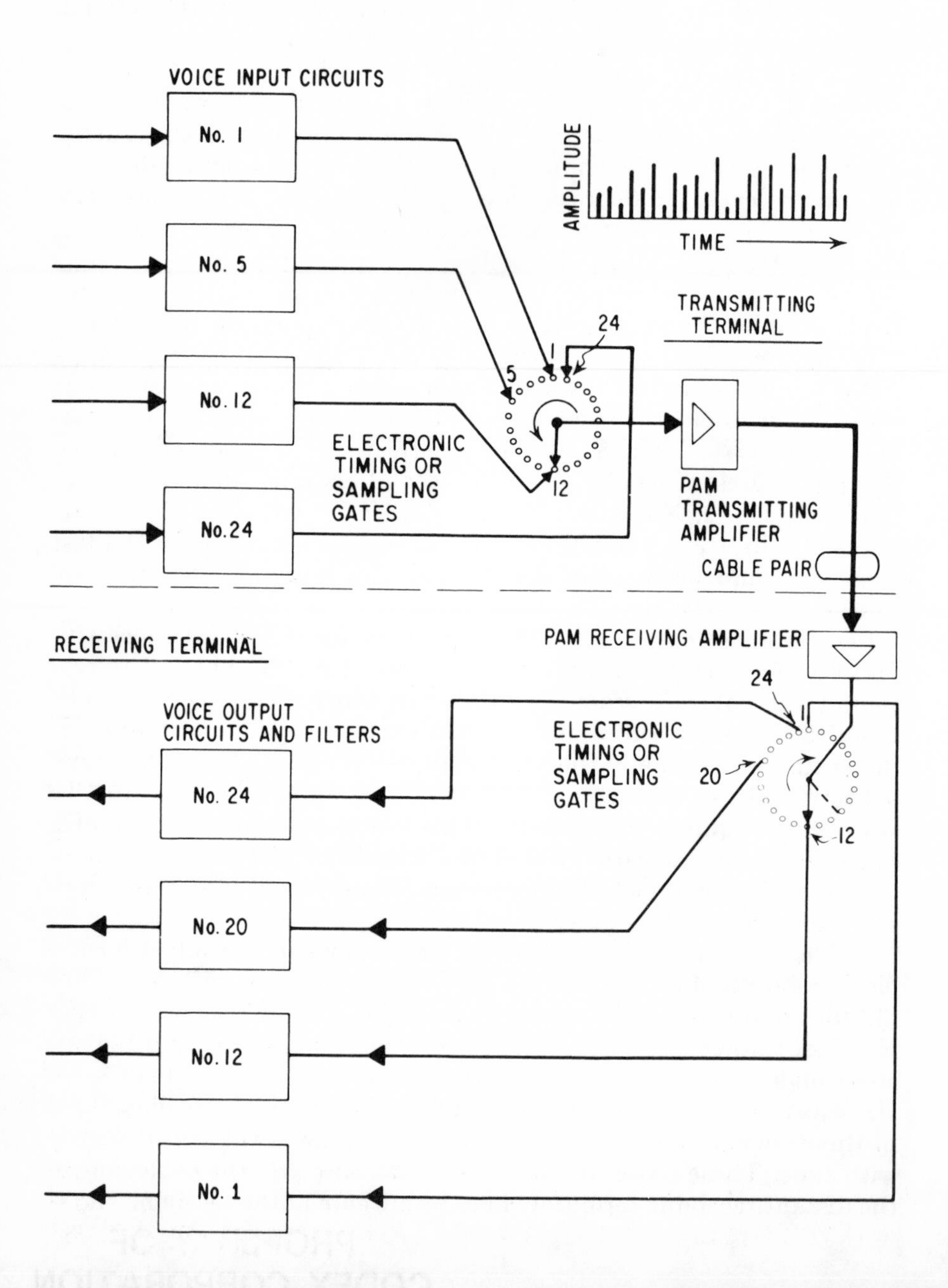

Pulse Amplitude Modulation (PAM)

The sampling operations that have been described are fundamental elements of TDM carrier systems. You will recollect that, in the FDM system, filters were necessary to separate each voice channel. Frequency translation was employed to scramble and unscramble (that is, modulate and demodulate) the various voice signals. In contrast, TDM systems utilize sampling pulses, timing circuits, and related electronic means for generating carrier channels. Recent improvements in solid state techniques have made the TDM and its related PAM methods more economical and reliable than the FDM.

In the elementary time division system previously described, each pulse is proportional to the amplitude of the signal represented. This method is appropriately termed *Pulse Amplitude Modulation (PAM).* A PAM carrier system, however, requires repeaters and amplifiers that possess excellent linearity characteristics. Otherwise, the resultant signal will be distorted whenever a high-level sample does not receive the same amplification that a low-level pulse does. Unlike the situation in FDM carriers, moreover, nonlinearity will not introduce crosstalk between channels. This is because the amplifier handles the various signal samples at different instants of time. As a result of this adverse trait and the greater bandwidth required, the pulse amplitude modulation method has not been employed for carrier systems.

Pulse Code Modulation (PCM) Fundamentals

The undesirable amplitude variations of pulse amplitude modulation can be overcome in a rudimentary fashion by converting each pulse sample to a binary-type pulse code. For example, a sequence of pulses, each of the same amplitude, can represent a particular voltage of a voice signal, say 0.30 volt. A different sequence of pulses can be used to describe a signal of 0.35 volt. Still another pulse sequence could portray a 0.40-volt signal, and the like. This pulse coding technique forms the basis of *Pulse Code Modulation (PCM).*

One can correctly assert that it is not efficient to send a number of pulses, instead of one pulse, for each sample of a voice signal. This technique, however, actually simplifies the requirements of a pulse code modulation carrier system. The reason is that in return for sending more pulses, it is only necessary to detect at any given time whether or not a pulse is present. This requirement is less difficult to satisfy than to have to determine the amplitude of a pulse. Moreover, the pulses will not be degraded by noise, crosstalk, or distortion in the cable or transmission medium as is the case with amplitude pulses.

The drawing on the following page illustrates a typical pulse train of a 24-channel *Pulse Code Modulation (PCM) Carrier System* such as the GTE Lenkurt 9001B, ITT type T124, Vicom T-1, and the West-

Pulse Code Modulation (PCM) Fundamentals (cont.)

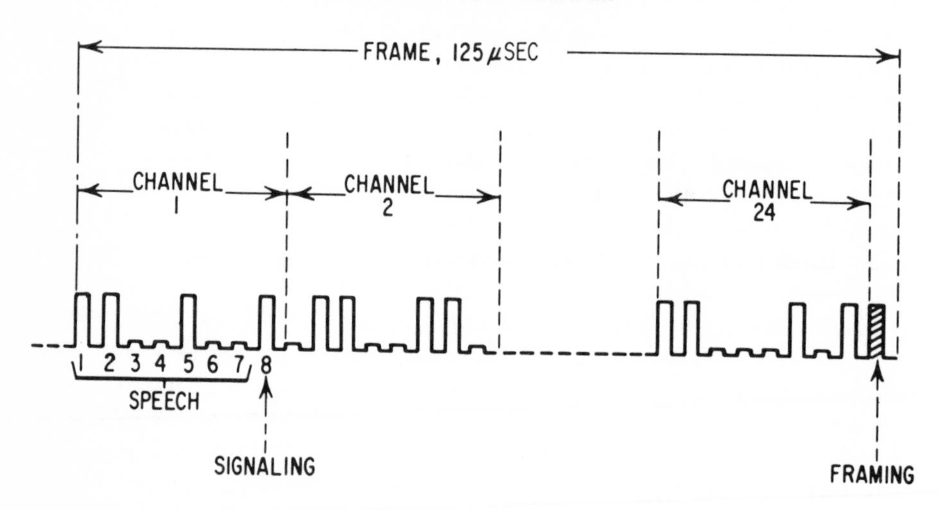

ern Electric T1. Each channel in the pulse train is initially sampled so as to produce a time sequence of 24 pulses, one pulse for each of the 24 channels. We will study how the pulse coding technique is utilized in these PCM carrier systems. In this connection, the fundamental operations consist of *sampling, quantizing,* and *encoding* the amplitude of a channel's sample into a *binary digit code* for *Time Division Multiplex (TDM)* transmission over cable pairs. These basic steps are shown in the following diagram.

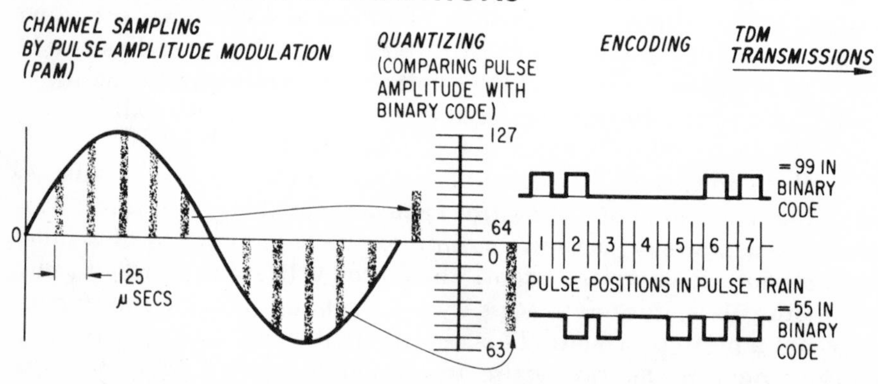

The speech signals in each voice channel are first sampled at the 8,000-Hz rate, which is at least twice as high as the highest speech frequency originated. This 8,000-Hz rate provides 125 microseconds spacing between the samples in a channel, and all 24 channels must

Pulse Code Modulation (PCM) Fundamentals (cont.)

be sampled and processed within this time period. The sampling operation, which portrays the analog waveform of the voice signal by a series of pulses, is termed *Pulse Amplitude Modulation (PAM)*. Note that the amplitude of each pulse is proportional to the instantaneous value of the speech signal in the respective voice channel. A chain of electronic gates (one for each channel), which are governed by precise timing circuits, controls the sampling process. The combined pulse train of these PAM samples for the 24 channels constitute a single PAM *frame*. The 24 channels, therefore, will produce 192,000 (24 × 8,000 Hz) individual samples per second for processing into PCM signals. Each channel's sampling slot will be 5.2 microseconds wide (125 microseconds/24).

The PCM signal is constructed from a unique *quantize* scale, using the *binary number system* for identifying the generated pulses. An electronic *encoder* circuit is used to compare the amplitude of each channel's PAM signal to a *binary code* scale comprising 128 values or levels. This scale is divided into 64 negative polarity levels, numbered 0-63; and 64 positive polarity levels, designated 64-127. Levels No. 0 and No. 64 have zero values and serve as the initial points on the binary code scale for establishing the respective negative or positive polarity values of the amplitude of each PAM sample received.

The *quantize* method can be considered as a comparison process because the *encoder* circuit compares the amplitude of a particular sample with the *binary code* scale, and assigns to it the value of the nearest one of the 128 possible levels. If the sample has a positive polarity, levels No. 64 to No. 127 would be used for the comparison and code assignment. For negative polarity samples, scale levels No. 0 to No. 63 would be utilized for these purposes. The *quantized* sample is next converted by the *encoder* circuit into a *binary code word* containing seven *binary digits*, usually abbreviated bits. The illustration on the next page depicts a typical *binary coded decimal* format used for *quantizing* and *encoding* operations in a PCM carrier terminal.

Binary Number System

The *binary digit* or bit refers to the digits in the *binary number system*. An explanation of the concept and operation of this system is beyond the scope and space limitations of this book. Therefore, only a brief synopsis is given to explain its use in PCM carrier systems. The binary system employs only two digits, *0* and *1*. It functions on the mathematical powers of 2 (2^n), in contrast to the ten digits and 10^n powers in the decimal system. A binary digit actually represents one of two possible conditions or logic states, such as on or off, high or low potential, positive or negative polarity, and the presence or absence of a pulse.

Binary Number System (cont.)

TYPICAL BINARY CODED DECIMAL TABLE FOR QUANTIZING AND ENCODING AMPLITUDE SAMPLES INTO PCM SIGNALS

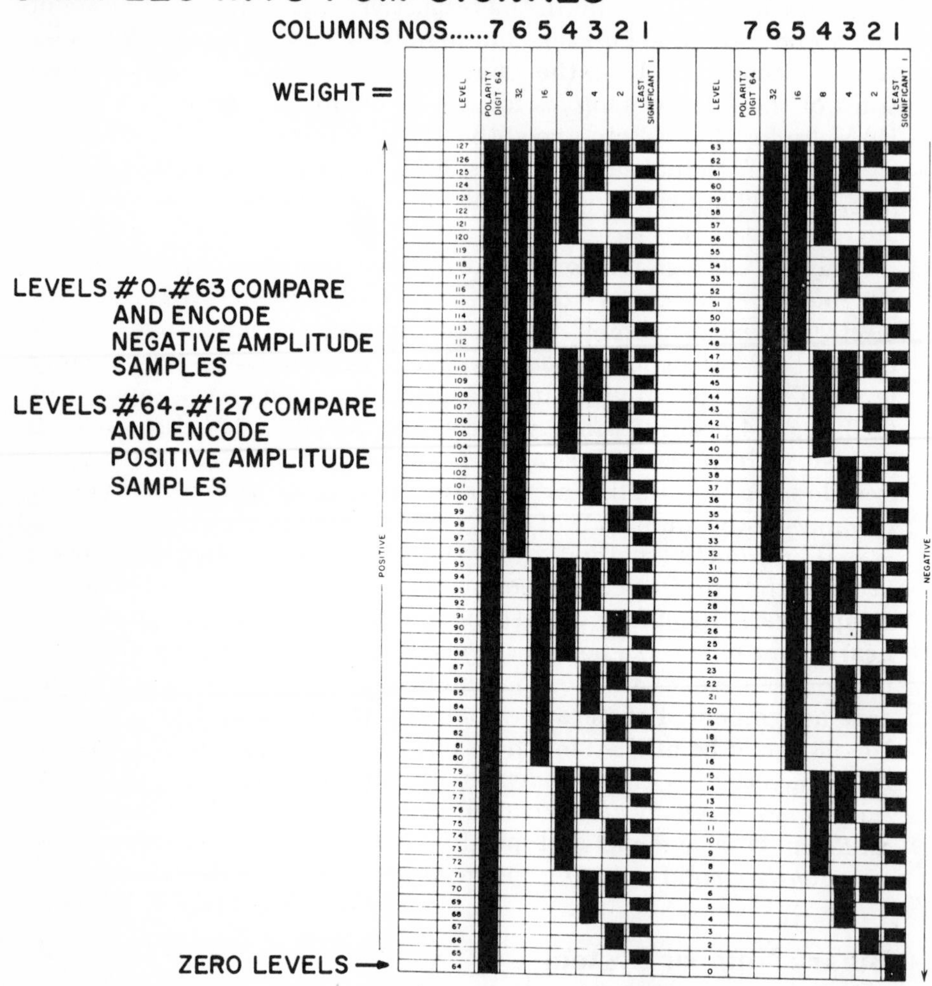

(Courtesy of ITT Telecommunications Div.)

Decimal numbers may be converted to the binary form and vice versa by the use of *binary coded decimal tables,* similar to the *binary code level* chart in the drawing. This chart employs a 7-bit code to provide 128 (2^7) possible quantizing levels. Each of the seven columns in this chart (numbered from right to left) has a value or weight based on the mathematical powers of 2, except that the first column on the extreme right always has a value of 1. Note that the values or weights, reading from right to left, are: 1,2,4,8,16, 32 and 64. Column No. 7 indicates the polarity of the particular channel's sample by a blank area

Binary Number System (cont.)

in levels 0-63 for negative polarity (representing a 0 bit), and a black area in levels 64-127 for positive polarity (representing a 1 bit). The black area in each column denotes a binary 1, and the blank space indicates a binary 0. For example, level No. 127 has all black areas in its columns which relate to binary digits 1111111. Referring to the weights in each column, the equivalent decimal number is:

$$64+32+16+8+4+2+1 = 127$$

Similarly, level No. 52 corresponds to binary digits 0110100 because its respective column weights are:

$$0+32+16+0+4+0+0 = 52$$

To provide for the binary coding, each channel's sampling slot of 5.2 microseconds width is divided into eight time slots, as shown in the diagram. The first slot represents the negative or positive polarity of the channel's amplitude sample. The next six time slots are for the encoded binary digits of the speech code that were derived from the *quantizing* process. The last or eighth slot is reserved for the signaling pulse of the particular channel. In this respect, an *on-hook* condition correlates to binary digit 0, and the *off-hook* state corresponds to binary digit 1.

Since binary digit 0 signifies the absence of a pulse, only pulses for binary digit 1 are transmitted over the cable pairs. We can say, therefore, that the presence or absence of individual pulses in the PCM pulse train, or binary *code word,* will represent any one of the possible 126 *quantizied* values that were used to encode the amplitude of the channel's sample. Levels No. 0 and No. 64, which have a zero value, are not used.

TIME SLOTS USED FOR TRANSMITTING BINARY WORD TO DENOTE AMPLITUDE OF CHANNEL'S SAMPLE AND CHANNEL'S SIGNALING STATUS

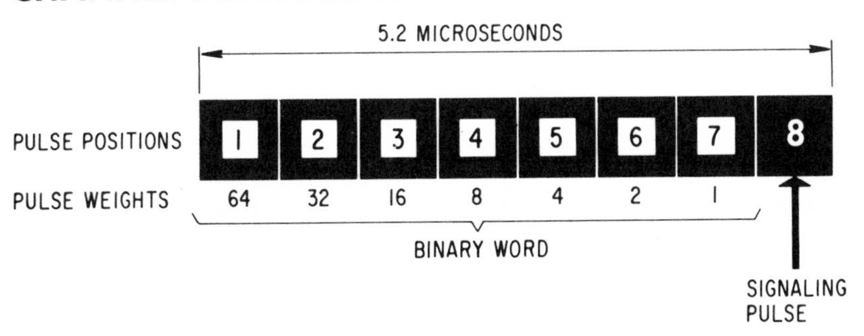

Composition of PCM Signal

A simplified version of the steps establishing a PCM carrier signal are illustrated in the figures below. Figure (A) represents an amplitude sample of a speech waveform in Channel No. 1 of a 24-channel PCM carrier system. This sample, as explained before, is the result of *pulse amplitude modulation (PAM).* Figure (B) shows the PAM frame formed by the amplitude sample in Figure (A), together with succeeding samples from the 23 other channels. Figure (C) depicts the PCM pulse train in Channel No. 1 resulting from the *quantizing* and *encoding* processes. The original amplitude sample in Fig. (A) has now been transformed into a train of 4 unipolar pulses (pulses of the same polarity and amplitude). Their positions in the pulse train (No. 1,2,5, 7) represent in binary digits (bits) the original speech sample in Fig. (A). The eighth or last pulse position in each train, described previously, is reserved for the channel's signaling information.

The aforementioned sampling, quantizing, and encoding processes for all 24 channels are completed within 125 microseconds and provide one *frame* of 24 8-bit binary words. An additional bit is included at the end of the Channel No. 24 pulse train for synchronization purposes. Thus, a complete PCM frame comprises seven binary digits for the speech code and one signaling bit for each of the 24 channels; added to this is one framing bit for a total of 193 $[(8 \times 24) + 1]$ unipolar bits. This number of bits multiplied by the sampling rate of 8,000 Hz yields the basic pulse repetition rate of 1,544,000 per second, or a bandwidth of 1,544 kHz for these unipolar pulse trains.

In order to reduce spectrum space, remove the dc component from the pulse train, and aid error detection at the receiving terminal, the unipolar pulse train is converted to *bipolar* form before transmission over the cable pair. In *bipolar* transmission, successive pulses in the train have opposite polarity. This helps detect errors because two successive pulses of the same polarity indicate error possibility in the PCM signal. With *bipolar* pulses, the total bit rate of 1,544,000 is halved and the bandwidth required reduced to 772 kHz. Figure (D) shows bipolar pulses equivalent to those in Fig. (C) for channel 1.

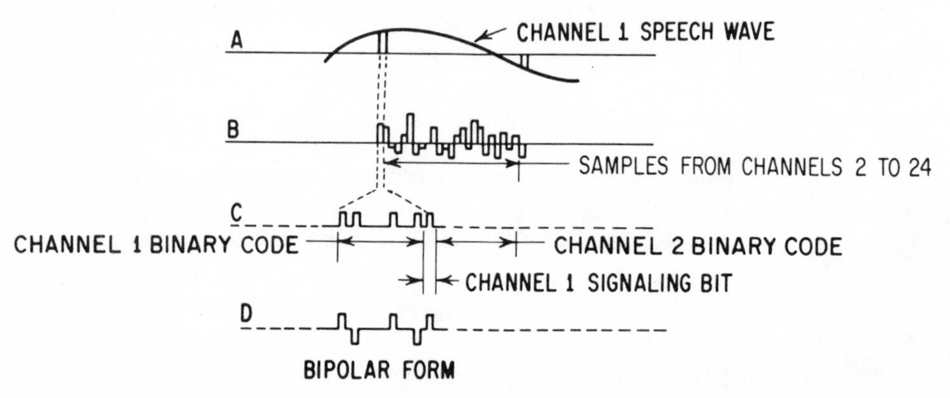

PCM Transmissions

The completed PCM signal is transmitted over one cable pair to the distant receiving terminal. A separate cable pair is employed for transmissions in the other direction. PCM carrier systems require regenerative type repeaters installed at about 6,000-foot (1.83 km) intervals, depending upon the wire gage of the cable pair. The design and operation of these regenerative repeaters will be described later.

At the PCM carrier receiving terminal, the incoming pulses first enter the receiving converter; and the bipolar pulses are changed to unipolar pulses. Next, the framing or timing pulses are recovered and fed to the *digit generator*. At the same time, the signaling pulses are removed from the pulse train. The signaling pulses are then directed to their respective channel signaling units, and the remaining unipolar binary pulses representing the speech code are sent to the *decoder*. The speech binary pulses are *decoded* to produce pulse amplitude modulation (PAM) samples, the amplitudes of which correspond to the quantized values that were initially established at the transmitting terminal. These PAM sample pulses are then expanded and amplified by the expandor circuit and distributed in turn to their appropriate channel units. A low-pass filter in each channel unit integrates these pulses to yield a replica of the original speech signal.

The value of the decoded sample at the receiving end, however, will not be exactly equal to the sample that was initially transmitted. This difference—which may be as much as one-half of a quantizing step—has been given the name *quantizing noise*. Unlike the idle circuit noise in FDM carrier systems, quantizing noise exists only in the presence of speech and has a tendency to be masked by it.

To minimize the effects of quantizing noise, a *compandor* unit may be employed in the transmitting terminal. This equipment has the property of making the quantizing steps small for small amplitude samples and large for correspondingly large samples. In the receiving terminal, this process is then reversed by an *expandor* unit.

The drawing on the next page is a simplified diagram of the elements of a PCM carrier system that we have already discussed. It is not a complete representation of an operational PCM carrier, such as the Western Electric T-1, that we will study in a subsequent section. The drawing includes a number of other elements that will also be discussed later in a great deal more detail—for example, the *master clock* and *framing generator*, the *digit generator*, and the *transmitting* and *receiving converters*.

The *master clock* unit controls the timing of the pulses that have been initiated by the *digit* and *framing generators*. The *transmitting converter* changes the unipolar pulses to bipolar pulses, whereas the *receiving converter* reverses this process. A careful examination of this drawing should enable us to understand the operations of the T1 PCM carrier system without any difficulty.

Types of PCM Carrier Systems

The growing demand for telephones and improved types of service increased the need for additional interoffice and toll-connecting trunks in cities and suburban areas. PCM carrier systems can provide

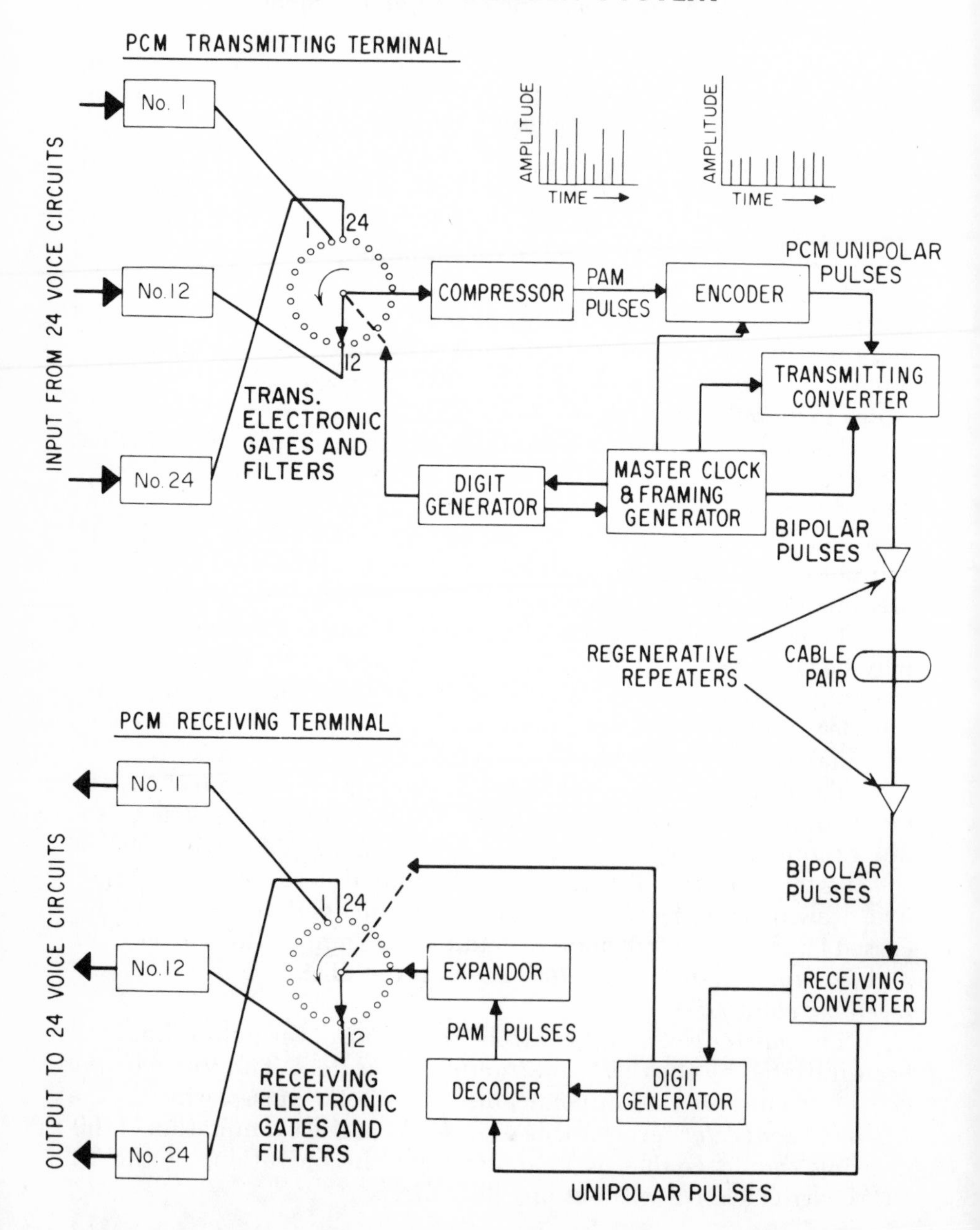

Types of PCM Carrier Systems (cont.)

these facilities over existing multiconductor cables more economically than installing new cables for voice-frequency circuits, or providing conventional FDM carrier systems. The Western Electric type T1 carrier was developed by the Bell Telephone Laboratories as the first practical PCM carrier system. It produced 24 voice channels over two cable pairs for interoffice and toll-connecting trunks. Regenerative repeaters are required about every 6,000 ft (1.83 km) when used over No. 22 gage cable pairs.

The terminal equipment of the T1 carrier includes the D1-type channel bank. This unit utilizes 128 quantizing steps or levels for evaluating and encoding the amplitude samples of the speech waveform into 7-bit *binary words,* as explained in the previous paragraphs. The number of quantizing steps is a measure of the transmission quality because it determines the *quantizing noise,* as described in the section PCM Transmissions. The resultant transmission quality and signaling features of the D1 channel bank, however, are not satisfactory for use in the DDD (direct-distance-dialing) or intertoll network of the Bell System.

The T2 PCM system, subsequently developed by the Bell Laboratories, supplies 96 voice channels that are suitable for use over the intertoll network. The D2-type channel bank used in the T2 system actually functions as four separate 24-channel PCM terminals. Every 24-channel unit, designated a *digroup,* connects to two cable pairs, one for each transmitting direction. The D3 channel bank was next evolved to provide the T1 system with the same equivalent transmission quality as the D2 type. Both D2 and D3 channel banks employ 256 (2^8) steps or levels, which necessitate the use of 8-bit *binary words* for encoding the voice samples.

Recall that all 24-voice channels must be sampled, quantized, and encoded in sequence within a 125 microsecond (1/8,000) period or *frame.* Therefore, each 8-bit *binary word* is permitted 5.2 microseconds for transmission. To minimize *quantizing noise* or distortion, all 8 bits should be utilized for encoding the amplitude samples. However, signal requirements for each channel (supervision and E&M signaling) must also be sent. Since signaling information is originated at a much lower frequency than voice, it can be sampled and transmitted at a slower rate. The signaling state, therefore, is sent in the eighth time slot of every sixth *frame* (frame No. 6, 12, 18 and 24, respectively). Thus, only 7 bits are used for encoding the voice samples in these frames, but all 8 bits are used in the other frames for voice encoding.

The latest and largest PCM system introduced by the Bell System is the T4M. It is designed to transmit PCM carrier signals over coaxial cables instead of the multiconductor cables used by the T1 and T2 systems. The T4M transmits 274 million bits per second, this is adequate for handling 4,032 voice channels over each of 18 coaxial cables

Types of PCM Carrier Systems (cont.)

in an ordinary cable conduit. A total of over 32,000 voice channels, therefore, can be conducted by a T4M system. Moreover, signals received at T4M terminals can be demultiplexed and sent over a number of T1 circuits or lines. This materially increases the flexibility of routing traffic in metropolitan areas. Regenerative repeaters or T4M line *regenerators* are spaced at approximately one-mile (1.6 km) intervals along the coaxial cable. The T4M system can handle voice, data, video, and other services as shown below.

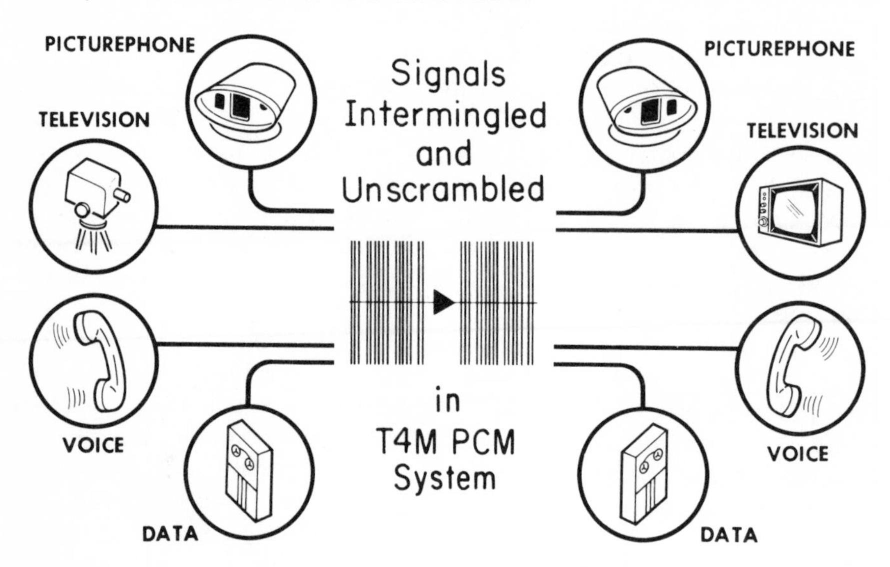

T1 Carrier Transmissions

The block diagrams on succeeding pages illustrate the operations of the D1 *channel bank* in a typical PCM carrier system, such as the Western Eleectric T1. Equivalent T1 PCM systems, including D1, D2, and D3 channel banks, are also produced by other manufacturers. This equipment is usually compatible with and can interface the Western Electric T1 type. A detailed explanation of this T1 system, accordingly, should help you to understand the operations of other PCM systems.

Let us follow the progress of a speech signal originated in voice channel No. 7, in the transmitting section of the D1 channel bank. The 24-channel units are of the plug-in type, and a *channel unit* connects with each incoming voice circuit. Six channel units connect to each of four *transmitting electronic gates* and *low-pass filter units.* Each of these units contains six sampling electronic gates and six low-pass filters arranged in two pairs—one pair for each group of 12 channels. Our channel No. 7 is served by the second gate and filter unit of Group 1.

TRANSMITTING SECTION OF D1 CHANNEL BANK OF THE T1 PCM CARRIER TERMINAL

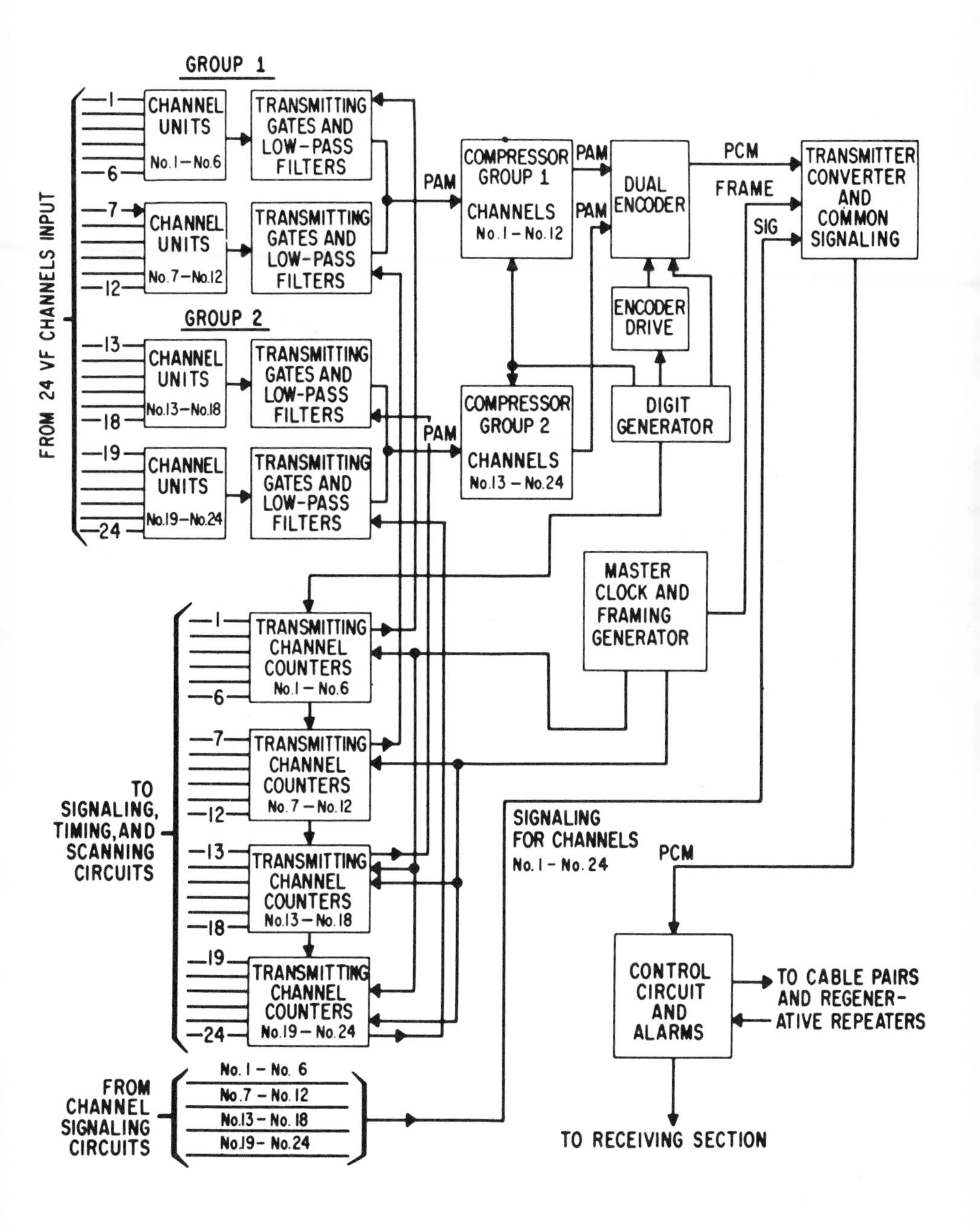

T1 Carrier Transmissions (cont.)

Four *transmitting channel counters* are shown in the lower portion of the drawing. These counters are under control of the *master clock and framing generator* circuit and provide the timing for each transmitting gate to sample the speech signals at a rate of 8,000 times per second. They also furnish the timing pulses for the *signaling timing and scanning* circuits associated with each of the 24 voice circuits. The sampling times of Groups 1 and 2 are interleaved so that channels No. 1 to 12 in Group 1 are sampled at odd-numbered times and No. 13 to 24 in Group 2 at even-numbered times. For example, the PAM pulse train will have channels in the sequence of 1,13; 2,14; 3,15; 4,16; 5,17; 6,18; 7,19; 8,20; 9,21; 10,22; 11,23; and 12,24.

The PAM output of each group of twelve transmitting gate circuits connects to a *compressor* circuit. The compressor functions almost instantaneously to reduce the wide range of the input PAM amplitudes to a very small range of output amplitudes in a nearly logarithmic relationship. As an example, a 60-dB input range (1,000-to-1 voltage ratio) would be reduced to about 36 dB (63-to-1 voltage ratio). In effect, the output voltage amplitude variations will be approximately proportional to the input variations in decibels. The output of the two compressors are directed to the *dual encoder* unit. PAM signals from channel 7 go through the compressor of Group 1 to the encoder.

The encoder circuit includes two summing amplifiers and comparison networks that are under the logic control of the *encoder drive* unit. The two PAM pulse trains from Groups 1 and 2 are encoded alternately into a single stream of PCM pulses. These pulses, which are unipolar (all have the same polarity), are next sent to the *transmitter converter and common signaling circuit.* Two other signals also connect to the transmitter converter circuit—namely, the signaling and framing pulses. The common signaling portion of this circuit receives a signaling pulse from each of the 24 channel units in turn, reshapes each one, and times it to interleave with the PCM pulses in the unipolar train. Since the PCM speech signal occupies seven of the eight pulse positions assigned to each channel sample, the signaling pulse will thus take the eighth pulse position. The third and last signal entering the transmitter converter unit is the framing signal from the framing generator in the master clock circuit. It will occupy a single pulse position per frame as a means of synchronizing the pulses at the receiving terminal.

The final step in the transmitter control unit is the transformation of the PCM unipolar pulses to a train of bipolar pulses. This is accomplished in the following manner. First, each pulse in the PCM pulse train of channel No. 7, for instance, is regenerated. Then, alternate pulses, whenever they appear, are inverted to form a bipolar sig-

T1 Carrier Transmissions (cont.)

nal. Lastly, the resulting PCM bipolar signal is transmitted through the control and alarm circuit over the cable pair to the distant receiving terminal.

RECEIVING SECTION OF D1 CHANNEL BANK OF THE T1 PCM CARRIER TERMINAL

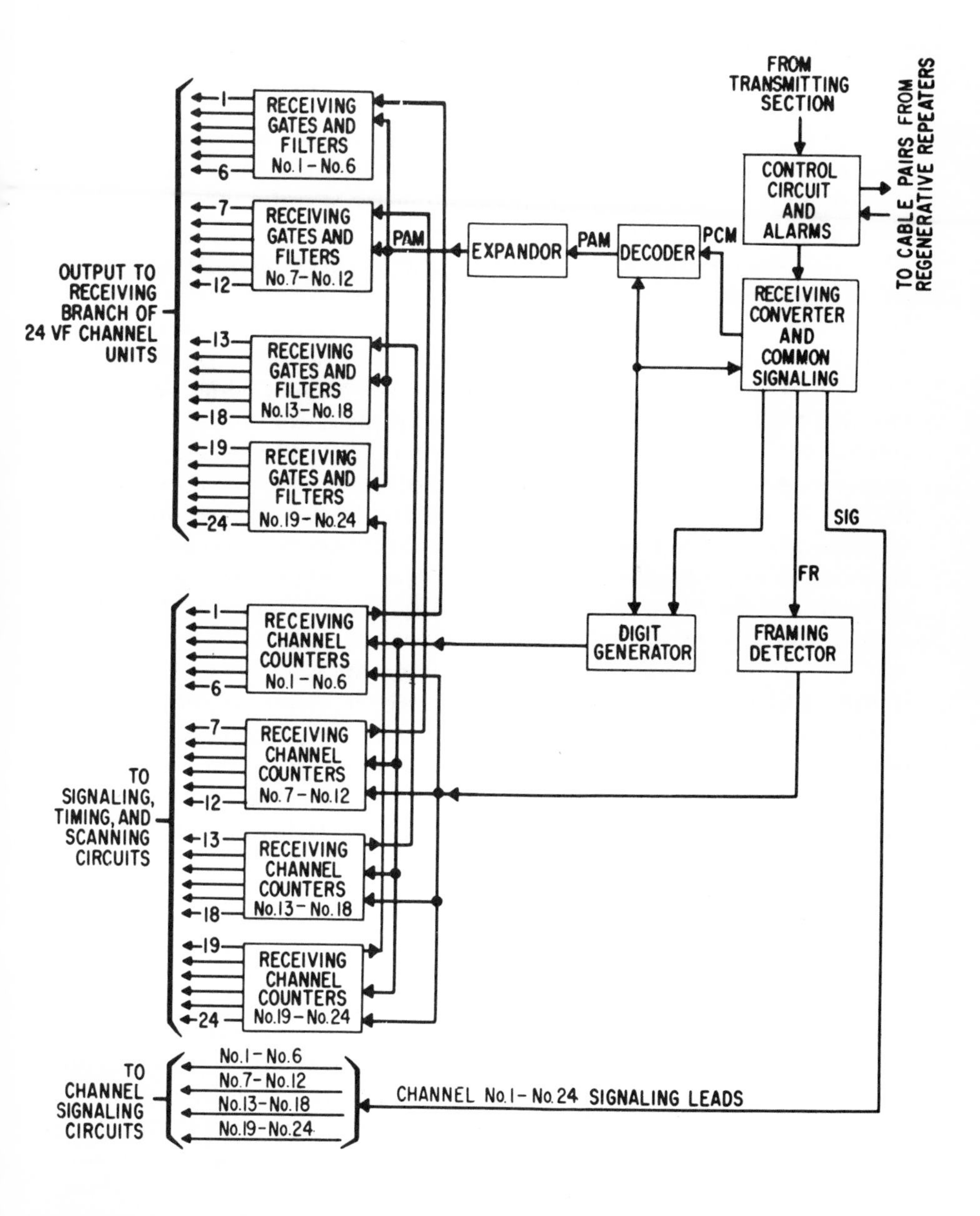

Receiving T1 PCM Carrier Signals

The incoming PCM pulses, on arriving at the receiving section of the T1 carrier terminal, first go through the *control and alarm* circuit to the *receiving converter and common signaling* unit. The bipolar PCM pulse train is then reconverted to unipolar form and regenerated by the *receiver converter* circuit. Next, the framing, signaling, and PCM speech pulses are simultaneously sorted out. The framing pulse, which had been inserted at the end of the pulse train of channel No. 24, is sent to the *framing detector* unit. The signaling pulses, which occupy the eighth position in each sample, are also extracted at this time for amplification and subsequent handling. The PCM speech pulses proceed to the *decoder* circuit for processing.

The *decoder* scans the seven pulse positions allocated to each speech sample. It combines the various binary code elements to form a sampling amplitude for the corresponding PAM pulse. The resulting PAM pulse train is then sent to the *expandor* circuit, which has the reverse characteristics of the *compressor* in the T1 transmitting terminal. As a result, PAM pulses are restored to their original uncompressed amplitudes before being applied to their respective *receiving gate and filter circuits*, as diagrammed on the previous page.

There are four groups of *receiving gates,* each of which serves six channel units. Channel No. 7 in our example is connected to the receiving gate associated with channels No. 7-12. The receiving gates operate one at a time in rotation under the control of the corresponding *receiving channel counters* shown in the lower portion of the drawing. Each PAM pulse is routed by its respective receiving gate through an individual low-pass filter to the receiving branch of its associated channel unit. The low-pass filter then integrates the PAM samples in order to yield the original speech signal.

Recall that the *receiving converter* circuit had extracted the signaling pulse related with each seven-pulse code. After amplification to a suitable pulse amplitude and duration, this signaling pulse (as with channel No.7)is passed along to its corresponding receiving signal gate in channel unit No. 7. This reproduced signaling state will be analogous to that originally scanned at the T1 transmitting terminal.

T1 Carrier Timing and Control Methods

We will now turn our attention to the ways and means used in order to time and to control the various circuits in the T1 PCM carrier system. The timing for the circuits in the T1 transmitting terminal originates from a crystal-controlled oscillator in the master clock and framing generator circuit. This oscillator output drives a *digit generator,* which is basically a series of blocking oscillators that provide a pulse sequentially on each of nine outputs. Eight outputs are employed to send out one of eight successive digits for use as needed for

T1 Carrier Timing and Control Methods (cont.)

encoding and other timing functions. One lead from each digit output is utilized for pulses of one polarity, and a second lead per digit is used for digit pulses of the opposite polarity. A ninth pulse is provided at the end of each frame for framing control in conjunction with the framing generator in the master clock unit. The schematic diagram on the next page shows the digit generator in a simplified form.

The pulses from the digit generator in the transmitting terminal drive the transmitting channel counters, which furnish the timing for the voice sampling and signaling scanning gates. The channel counter stages, like the digit generator, consist of blocking oscillators turned on in order by one digit pulse and turned off by another pulse.

At the receiving terminal, the timing for the various circuits is very similar to that which has been described for the transmitting terminal. The master timing or clock signal, however, is derived from the incoming pulse train itself instead of from a crystal-controlled oscillator. This clock signal is produced in the receiving converter circuit as part of the pulse regeneration process. It also drives a digit generator unit that duplicates the one in the transmitting terminal.

As is also the case in the transmitting terminal, the pulses from the digit generator drive the receiving channel counters, which time the receiving gates and filter circuits and the channel signaling timing and scanning circuits, and a framing pulse is produced at the end of each frame as determined by the condition of the receiving channel counters.

In summary, the pulse code modulation bit rate, digit pulse frequency, and the channel and frame ratios in the receiving terminal are identical with those that originated in the transmitting terminal. Therefore, once synchronism has been achieved, it can be maintained indefinitely just so long as the incoming pulse train is not interrupted.

In the eventuality such an interruption occurs, phase synchronism or framing will be restored under control of the framing detector circuit. The detector circuit receives the framing signal that is generated in the receiving timing circuits and compares it with the corresponding signal from the incoming pulse train. It can introduce an additional pulse per frame until full synchronism is again attained.

PCM Regenerative Repeater

Regenerative type repeaters are used in PCM carrier systems such as the Western Electric T1. *Regenerative* means that the PCM pulse train is reshaped and retimed. Reshaping may be defined as the raising of the level of a pulse to the point at which a pulse versus no-pulse decision can be made readily. Pulse retiming is the sending out of a new pulse by the repeater to replace an incoming signal pulse.

The diagrams on page 178 show the essential configuration of a typical PCM regenerative repeater used for two-way transmissions. It

SIMPLIFIED SCHEMATIC OF DIGIT GENERATOR CIRCUIT OF T1 PCM CARRIER

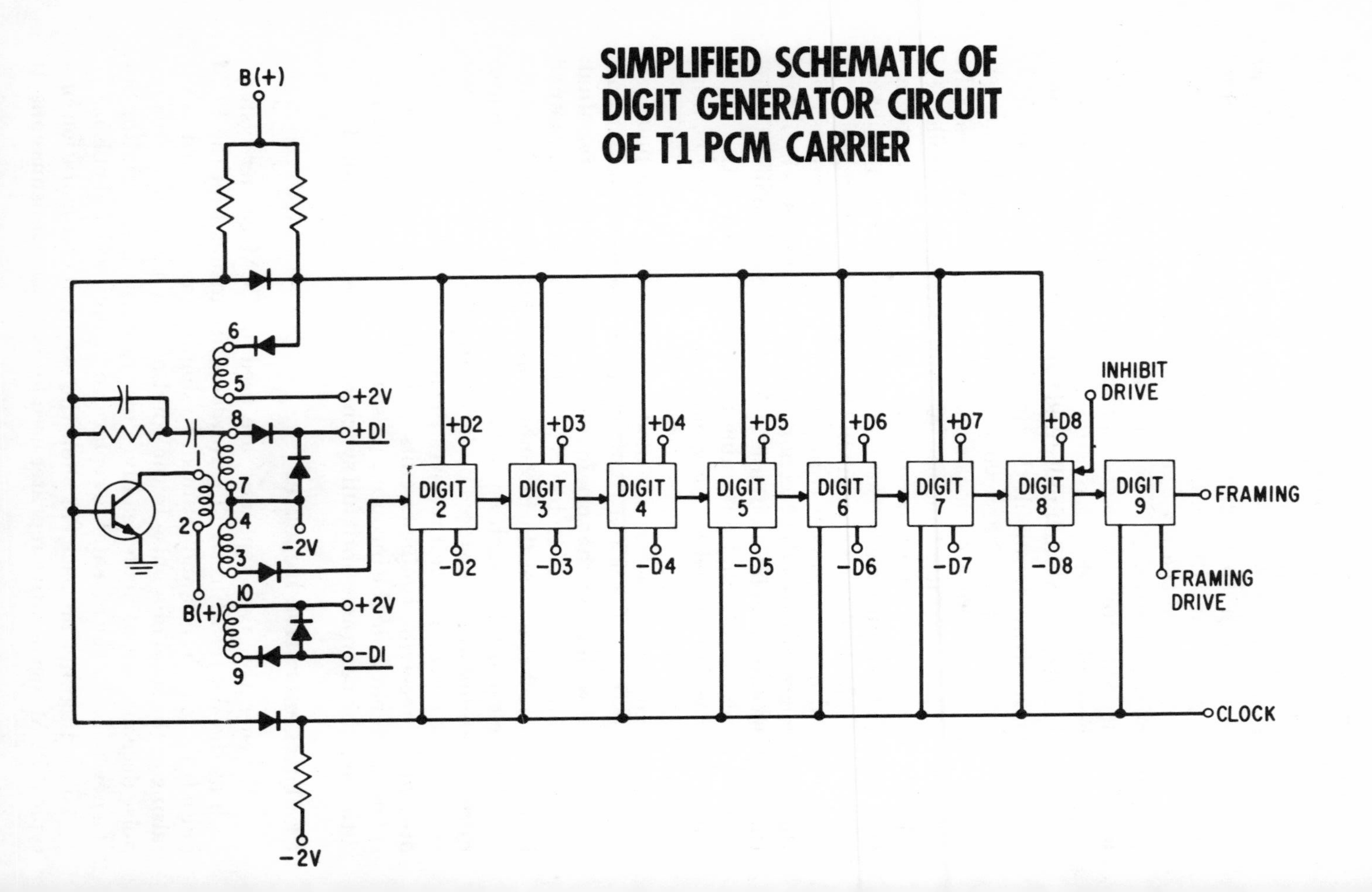

PCM Regenerative Repeater (cont.)

is comprised of two *line build-out* networks, two *regenerators*, and a *power circuit.* The power circuit provides a regulated voltage for the operation of the transistors and the other solid-state elements in the repeater. Provisions are included for either looping the line power or connecting it through in whatever way may be required.

A *regenerator* is used for each direction of transmission. Each regenerator contains a preamplifier, a clock rectifier, and processing circuit, timing gates, and a balanced pulse regenerator. The schematic diagram on the following page illustrates these elements.

The incoming bipolar PCM signal is initially amplified by the preamplifier circuit. This action also tends to reshape each pulse, thereby reducing its dispersion into adjacent time slots. The amplified signal next drives the threshold bias, clock processing, and pulse regenerator circuits. The threshold bias circuit sets the pulse decision level for the signal. This determines for each time slot whether a pulse is to be regenerated.

The clock rectifier converts the incoming bipolar PCM signal into a unipolar pulse train. The resulting pulse train will contain a high component of the 1,544-kHz pulse repetition voltage that was originally generated in the transmitting terminal. This 1,544-kHz signal is amplified and shaped or equalized in the clock circuit in order to provide *on* and *off* timing pulses for the beginning and end, respectively, of each pulse position.

There are two timing gates, as is revealed by the drawing—one gate for each signal polarity. As a result, whenever an incoming pulse of either polarity coincides in time with the "on" timing pulse, the pulse generator of the same polarity will be triggered to emit a new pulse. The timing gates and pulse generators will normally operate alternately on incoming pulses that are error-free. The line build-out units provide attenuation pads that will compensate, in whatever way required, for any variations in the resistance and capacity of the cable conductors.

PCM Repeater Spacings

Regenerative repeaters are normally powered by 48- or 130-volt batteries in the central offices at each end. This dc is sent over a simplex loop consisting of the two cable pairs feeding through each of the repeaters (see drawings on next page). A total of six cable or line repeaters may thus be powered from one central office. The dc along the cable pairs also serves as sealing current for unsoldered splices on interoffice trunk cables. Since line current between terminals is maintained at about 0.14 ampere, nominal voltage drop in each repeater is about 10 volts. Voltage drop due to the cable conductors' resistance between repeaters (line section) may be of the same magnitude, depending upon the cable pairs' length and size.

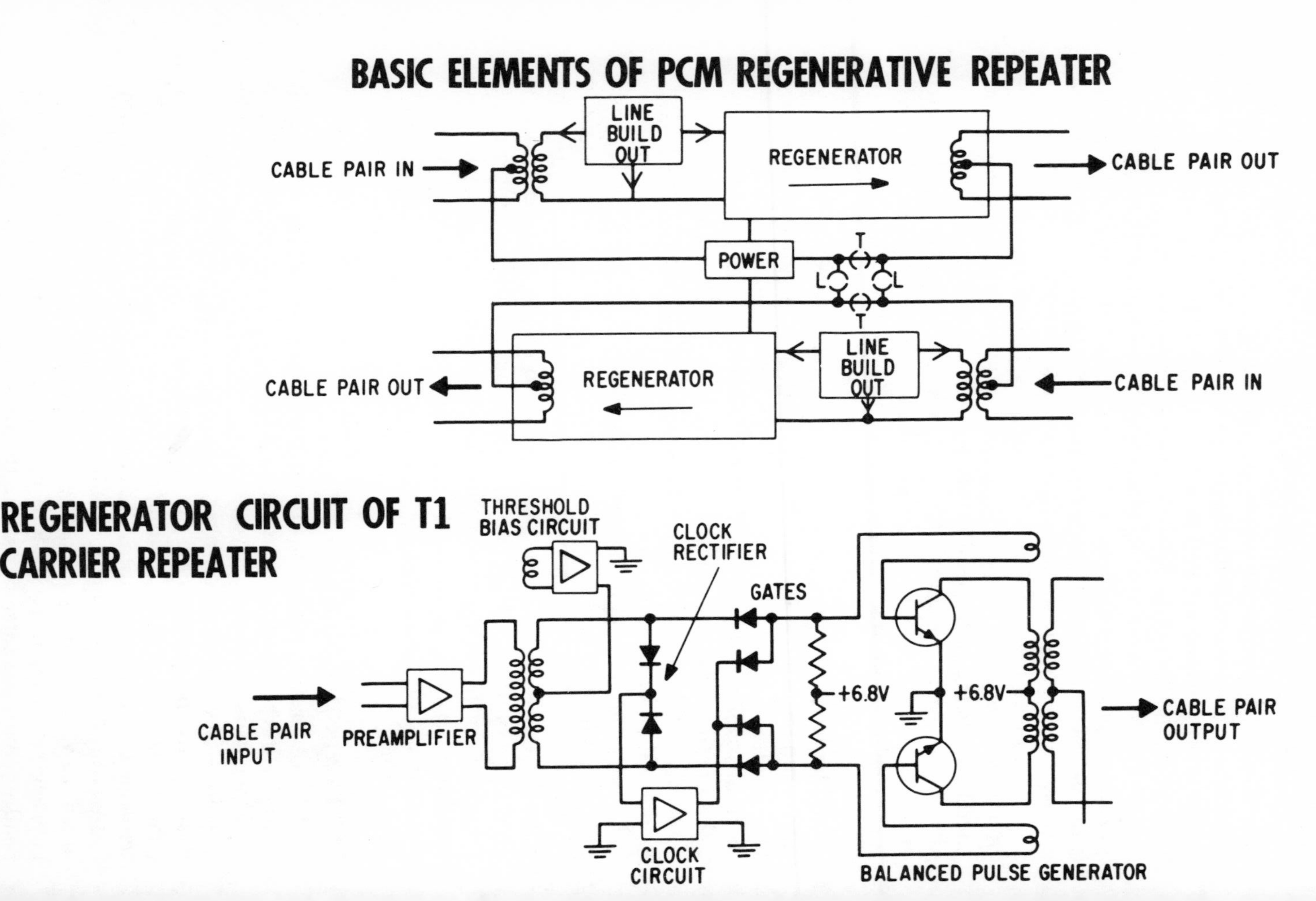
BASIC ELEMENTS OF PCM REGENERATIVE REPEATER
LINE BUILD OUT
CABLE PAIR IN
REGENERATOR
CABLE PAIR OUT
POWER
T
L
L
T
CABLE PAIR OUT
REGENERATOR
LINE BUILD OUT
CABLE PAIR IN
REGENERATOR CIRCUIT OF T1 CARRIER REPEATER
THRESHOLD BIAS CIRCUIT
CLOCK RECTIFIER
GATES
+6.8V
+6.8V
CABLE PAIR INPUT
PREAMPLIFIER
CABLE PAIR OUTPUT
CLOCK CIRCUIT
BALANCED PULSE GENERATOR

PCM Repeater Spacings (cont.)

Repeater spacing distances are related to cable attenuation at the highest frequency of the transmitted PCM bandwidth. For T1 carriers, this frequency may be taken as 772 kHz. Repeater separations may thus be expressed in feet in accordance with the loss of 772 kHz for the multiconductor cables normally installed. The table lists typical cables and respective spacings of regenerative repeaters in a T1 system. (The notation *PIC* refers to plastic-insulated conductors; other cables use paper-insulated conductors.)

Type of cable	*Loss per mile at 772 kHz*	*Carrier repeater spacings*
No. 19 (PIC)	15.3 dB	10,500 ft (3.2 km)
No. 19	19.2 dB	7,900 ft. (2.4 km)
No. 22	26.6 dB	6,000 ft. (1.83 km)
No. 24	29.6 dB	5,200 ft. (1.59 km)
No. 26	39.5 dB	3,800 ft. (1.16 km)

Questions and Problems

1. What is the fundamental difference between FDM and TDM carrier systems?
2. What sampling rate is required to faithfully represent a 2,500-Hz tone? Why? What sampling rate is generally used in PCM carrier systems?
3. What are the main advantages of PCM carrier systems?
4. What fundamental operations are performed in a PCM carrier system? What is the T1 carrier system's channel capacity? How may pulse positions comprise a pulse train for each channel?
5. Do all the pulse positions in a PCM pulse train carry speech samples? Explain. What is the time interval in the T1 carrier system for completing the processing of all channels? How many bits comprise a complete PCM frame? Explain.
6. What determines the bandwidth of the basic pulse repetition rate in the T1 carrier system? How is it reduced for transmission over telephone cables?
7. How was the transmission quality of the original T1 carrier system improved for use over the intertoll (DDD) network?
8. Name the principal circuits in the receiving section of a T1 carrier terminal. What function is performed by the Framing Detector?
9. How many regenerator circuits are provided in a PCM repeater? Name the basic elements of the regenerator.
10. What are the normal repeater spacings for a T1 carrier system over No. 19 gage PIC cable? Over No. 22 gage cable? What is the nominal loss per mile at the PCM frequency for these cables?

General Applications of Carrier Systems

The underlying or fundamental principles of analog (FDM) and digital (PCM) carrier telephone systems have been explained in this book. The basic equipment elements of single and multichannel carrier systems were described, the modulation and demodulation processes were discussed, the various types of filters and their characteristics were pointed out, and examples were given of simple filter designs applicable to FDM carrier systems. The general objective was to simplify the technical descriptions so that the reader could readily grasp the basic techniques of carrier telephony.

The carrier systems covered so far pertain to the telephone cable plant involving interoffice, toll-connecting, short-haul toll trunks, and subscriber lines. We will now consider broadband carrier systems used with coaxial cables, point-to-point microwave radio links, and communication satellite systems. These facilities provide the long-distance circuits for the DDD network and other intertoll operations within the continental U.S. and Canada. The carrier techniques that we have studied are also employed in submarine coaxial cables. These cables are equipped with repeaters and cross the oceans to provide global telecommunication services. Communication satellites and associated ground stations also employ these carrier techniques.

The chart lists important types of transmission facilities using carrier systems, and the range of their present voice-channel capacities. Note that any voice channel (250-3,450 Hz) in a carrier system may be multiplexed by carrier techniques, thereby furnishing up to

Type of Carrier Transmission Facility	*VF Channel Capacity per Carrier Facility Route*
A. Open-Wire Lines	2 to 16
B. Multiconductor Cables (per 4-wire circuit)	4 to 24
C. Coaxial Cables (L1 to L5 Carrier Systems)	600 to 108,000*
D. Point-to-Point VHF Radio Links	2 to 24
E. Point-to-Point Microwave Radio Systems	600 to 29,520
F. Submarine Repeater Cables	845 to 4,000
G. Communication Satellite Systems	240 to 6,250

*per pair of coaxial cables.

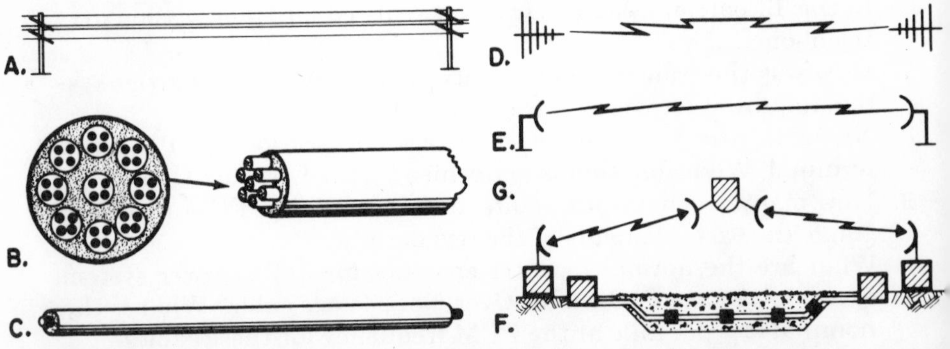

A. B. C. D. E. G. F.

General Applications of Carrier Systems (cont.)

about 20 teletypewriter channels, or used as a binary data channel to handle up to 2,400 bits per second (bps). The data channel capacity can be increased to 4,800 bps by providing a properly equalized or conditioned private line circuit between the dataphone set and the multiplex equipment in the toll central office.

Coaxial Cables

We have learned that cable carriers provide up to 24 one-way channels per cable pair. Inherent features of cables, noise, and other factors impose this limit. Although additional pairs may be loaded with channels up to the capacity of the particular cable, this arrangement cannot provide the thousands of high quality circuits needed for an intertoll network. Echo effects, phase delays, and noise adversely restrict the system's transmission range. Recent developments have made possible the use of broadband transmission facilities over coaxial cable conductors and microwave radio relay circuits to provide large numbers of high-quality channels at considerably reduced cost.

Coaxial cables in the Bell System's nationwide intertoll or DDD network are composed of from 18 to 22 copper tubes, each ⅜ in. (9.53 mm) in diameter, and arranged in a tight ring inside a lead and polyethylene cable sheath. For instance, coaxial cables used with the older type 3½ in. (8.9 cm) cable ducts under city streets have a capacity of 18 coaxial copper tubes. The newer 4 in. (10.2 cm) cable ducts can accommodate coaxial cables having 22 copper tubes. Direct-burial coaxial cables usually contain 20 tubes. Each copper tube in the coaxial cable is wrapped with steel tape for added protection and electrical shielding. A copper wire (normally No. 10 gage) is secured in the exact center of each copper tube by small plastic disc insulators, installed about one-inch (2.54 cm) apart, as shown in the drawing. The copper tubes prevent radiation of carrier frequencies so that identical carrier frequencies may be transmitted over individual copper tubes without mutual interference.

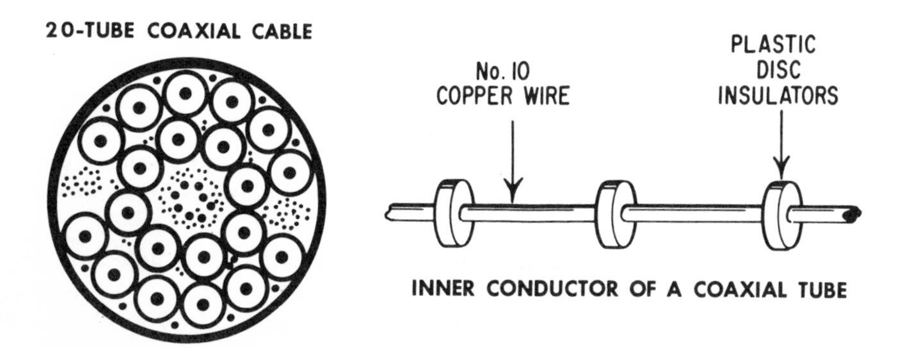

Coaxial Cables (cont.)

Two copper tubes comprise one carrier system; one tube for each direction of transmission. Therefore, the 18, 20, and 22 tube-coaxial cables can furnish 8, 9, and 10 carrier systems, respectively, because one pair of tubes is utilized as a carrier protection system. A transmission failure in any of the carrier systems of the particular coaxial cable will cause the message traffic in the effected system to be diverted to the protection system.

The drawing shows the composition of a 20-tube coaxial cable which is about 3 in. (7.6 cm) in diameter and weighs approximately 9 lb per foot. A number of plastic-insulated wire pairs included in the center of the coaxial cable are used for control, alarm, order-wire, and similar functions between attended maintenance centers. This coaxial cable, along with 18 and 22 tube types, are also employed in the transcontinental *hardened route* of the Bell System. These coaxial cables, their related main stations, repeaters, and power-feed stations are underground and can withstand most natural disasters, as well as a nuclear blast that is not a direct hit. Since coaxial cables are invariably buried, the relative constant temperature of the earth minimizes signal level variations. Dry air or nitrogen gas introduced under pressure keeps out moisture and helps to detect cable damage.

Broadband Carrier Systems

The original broadband carrier system suitable for coaxial cables is the Western Electric L1; this was developed by the Bell Telephone Laboratories and first installed in 1941. It is a single sideband suppressed carrier system utilizing the 60- to 2,540-kHz frequency band. A total of 600 voice channels can be transmitted over a pair of coaxial cables, one for each transmission direction. The modulation steps and related methods used in the L1 to assemble the 600-voice channels for carrier transmission formed the basis for the subsequent development of the much larger carrier systems now in use. These include the Western Electric types L3, L4, and L5, GTE Lenkurt 46A, 46C, and similar equipment made by ITT Corporation and others.

The fundamentals of multichannel carrier operations, described in the section on FDM Cable Carrier Systems, also apply to the aforementioned broadband carrier systems. The basic modulation plan of the initial L1 carrier, which is illustrated in the drawing, provides the *supergroup* framework for the *mastergroups* and *jumbogroups* concepts used in current coaxial carrier systems. In order to understand the operations of the Western Electric L3, L4, L5, and similar carrier systems of other manufacturers, let us follow the modulation steps and procedures involved in the L1 system.

The first modulation step assembles a group of 12 voice channels (each of 250-3,450-Hz bandwidth) into a *channel bank* or group occupying the 60-108-kHz range. This is accomplished by modulating

Broadband Carrier Systems (cont.)

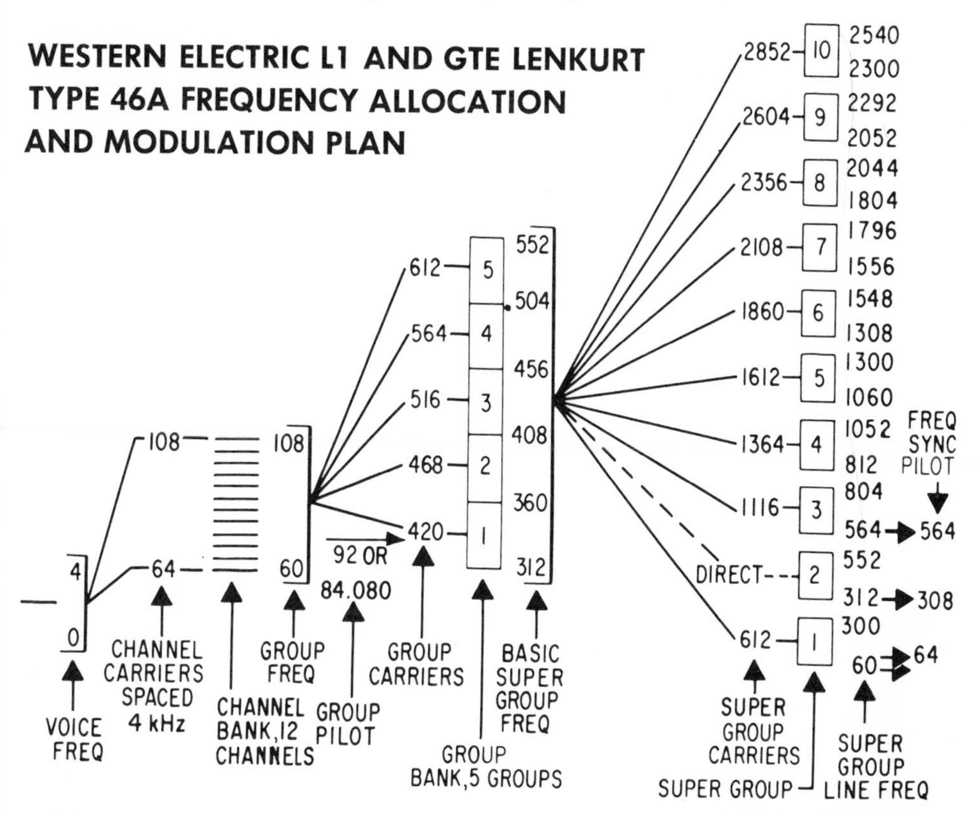

the 12-voice channels, using channel carriers spaced 4-kHz apart in the frequency range 64-108 kHz. Only the lower sideband products are used because the upper sidebands and related channel carriers are suppressed. This process is explained in the FDM Modulation and Demodulation Process section. Thus, the 600-voice channels are divided into 50 groups of 12 channel banks in the 60-108-Hz range.

The second modulation step shifts the series of five groups (60 voice channels) to the 312-552-kHz range. Group frequencies of 420, 468, 516, 564, and 612 kHz are employed for this purpose, as shown in the drawing. This combination of 60 voice channels is termed a *basic supergroup.* In a similar manner, the other channel groups are assembled to comprise a total of ten basic supergroups, each in the frequency band of 312-552 kHz.

The third or last modulation step places the ten supergroups into the allocated line frequency spectrum of 60-2,540 kHz. This is achieved by the nine supergroup carriers indicated in the drawing. Note that supergroup No. 2 is not modulated in the third step because it already occupies the 312-552-kHz band. Pilot frequencies of 64, 308, and 564 kHz are used for carrier synchronization and control purposes at the distant receiving terminal.

Broadband Carrier Systems (cont.)

The demodulation process at the carrier receiving terminal utilizes three similar steps of carrier frequencies, but in reverse order. Each of the ten supergroups is first demodulated into its respective five groups. Then, each group and its exact carrier frequency is demodulated into its twelve channel banks. The last step demodulates each channel bank into the voice-frequency band (250-3,450 Hz).

An important characteristic of carrier frequencies should be noted with respect to coaxial cables. Wire pairs in telephone cables are twisted and, consequently, are considered to be balanced with respect to ground. In coaxial cables, the copper tubes serve as conductors as well as shields for the carrier signals. Coaxial cables, therefore, are unbalanced to ground. Moreover, carrier currents are much higher in frequency than voice-frequency currents. Thus, they travel on the outer surface of the inner conductor and on the inner surface of the copper tube opposite the inner conductor; this is due to the *skin effect* phenomenon. This effect, however, diminishes with decrease in carrier frequency. As a result, lower carrier frequencies will appear on the outer surface of the copper tube. Since many copper tubes are paralleled in coaxial cables, severe crosstalk will result. For this reason, present coaxial cable systems normally do not employ carrier frequencies below about 300 kHz.

Coaxial Cable Carrier Systems

L1 Carrier System. The initial L1 coaxial cable carrier system used an 8-tube coaxial cable sheath. It provided 1,800 voice circuits, 600 circuits/copper tube pair times 6 tubes. One pair of copper tubes was used for the protection of the circuit. Unmanned repeaters installed in huts were spaced at 8-mile (12.9 km) intervals. With the ever-rising demands for long-distance telephone circuits after World War II, the L3 carrier system was developed in 1953 to replace the L1.

L3 Carrier System. The L3 carrier system, like the L1, employs vacuum tubes. It can provide 1,860 voice circuits (VF channels) for a distance of 4,000 miles (6,437 km), requiring over 1,000 repeaters in tandem. Moreover, the L3 has a capacity of one 4.2-MHz television channel, plus 660 VF channels. Unmanned repeaters are installed at 4-mile (6.4 km) distances. With the older type 12-tube coaxial cable, the L3 provides 9,300 VF channels (1,860 $\times$ 5 copper-tube pairs).

The master oscillator at the originating L3 transmitting terminal controls the generation of the line pilot frequencies indicated in the drawing. They are used as reference frequencies by the other terminals along the coaxial cable; thus insuring the extreme accuracy required of the various carrier frequencies generated within the L3 system. Pilot frequencies are used for repeater regulation and control.

The drawing is a simplified outline of the L3 frequency allocation plan, which utilizes the 308-8,320-kHz spectrum. Similar modu-

Coaxial Cable Carrier Systems (cont.)

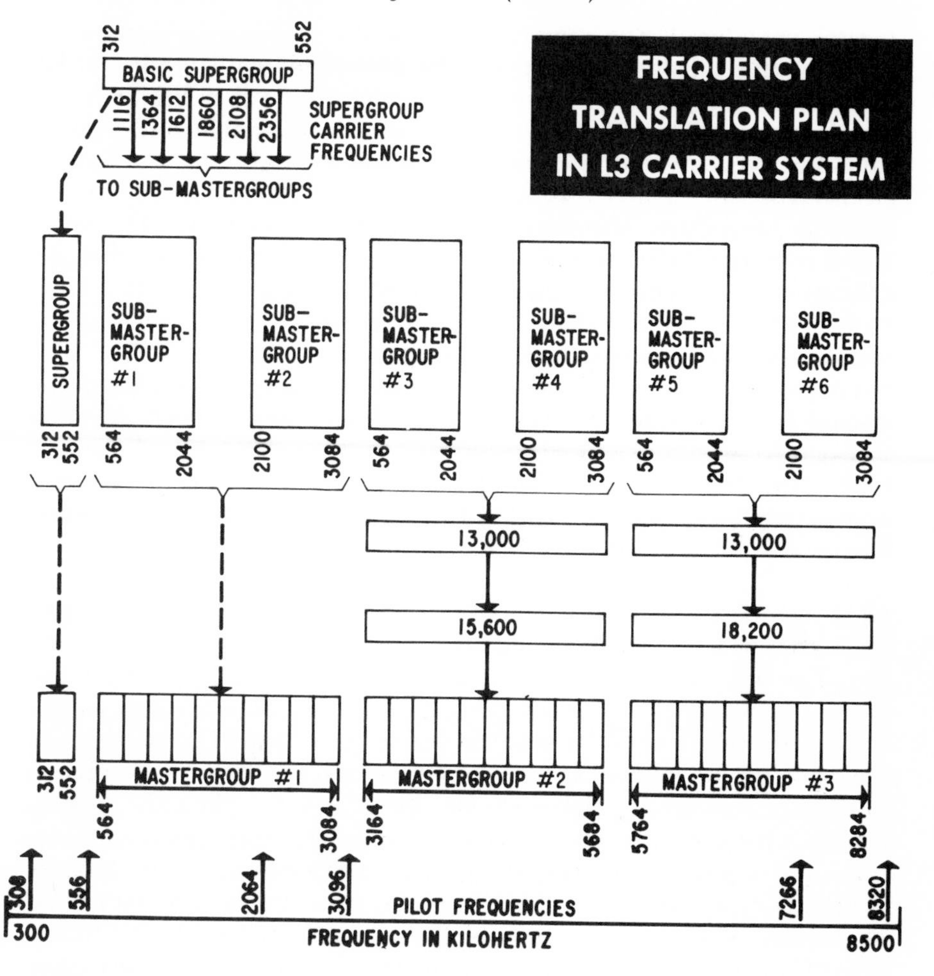

lation procedures, as described for the L1 system, are employed to place five groups (each of 12-VF channels) into a basic *supergroup* (60-VF channels) in the 312-552-kHz band. A total of 31 *supergroups* (1,860-VF channels) are evolved in this fashion. Thirty supergroups are next modulated by the six supergroup carrier frequencies, listed in the drawing, into three pairs of *sub-mastergroups.* Each odd-numbered *sub-mastergroup* contains 6 supergroups (360-VF channels), every even-numbered sub-mastergroup has 4 supergroups (240-VF channels), totaling 1,800 VF channels. Note that the first basic supergroup (312-552 kHz) is not further translated but is used as the lowest frequency segment in the L3 frequency plan. This arrangement increases the capacity of the L3 system to 1,860 voice channels.

The next modulation step combines the six *sub-mastergroups* into three *mastergroups.* Sub-mastergroups No. 1 and No. 2, which occupy the 564-3,084-kHz spectrum, are not modulated and become

Coaxial Cable Carrier Systems (cont.)

mastergroup No. 1. Sub-mastergroups No. 3 and No. 4 are first modulated by the 13,000-kHz carrier. The upper-sideband products and carrier are suppressed, and the lower sidebands are then further modulated by a 15,600-kHz carrier. The resultant lower sidebands (3,164-5,684 kHz) are utilized for *mastergroup* No. 2. Similarly, sub-mastergroups No. 5 and 6 are translated by carriers 13,000 and 18,200 kHz in sequence, and form *mastergroup* No. 3 in the 5,764-8,284-kHz range. The complete L3 spectrum, including pilot carriers occupies 308-8,320 kHz, as shown in the drawing on the previous page.

L4 Carrier System. Increasing requirements for toll and long distance facilities to meet growing demands for voice, data, and television circuits led to the development by the Bell System of the L4 coaxial carrier system. The L4, initially introduced in 1967, employs transistors and other semiconductor devices, in place of vacuum tubes. It has a capacity of 3,600-VF channels per pair of copper tubes in a coaxial cable sheath. The 20-tube hardened coaxial cable, therefore, can furnish 32,400 telephone channels, a route capacity greater than that provided by microwave radio systems. Repeaters are spaced every 2 miles (3.2 km) because of increased transmission loss due to the higher 564-17,548-kHz band occupied.

The basic *mastergroup* concept of the L3 carrier is also employed in the L4 frequency allocation plan. Six mastergroups are sequentially aligned in the frequency spectrum of 564 to 17,548 kHz, as shown in the drawing. Each mastergroup has a capacity of 10 supergroups, or 600 VF channels, as explained for the L3 system. The six mastergroups, which provide up to 3,600-VF channels, are termed a *jumbogroup,* as subsequently utilized in the L5 carrier system. The L4, like the L3, is designed for a 4,000-mile (6,437 km) circuit. The repeaters, which use solid-state components, are normally in manholes; the dc power for the repeaters is sent over the inner conductors of the copper tubes in the coaxial cable sheath.

L4 LINE FREQUENCY PLAN

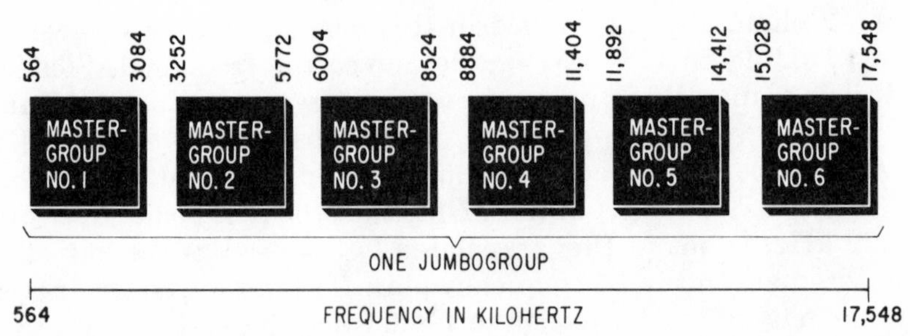

L5 Carrier System. The continued need for more toll and long distance circuits, particularly for high-speed data and video traffic,

Coaxial Cable Carrier Systems (cont.)

caused the Bell System to further improve its coaxial cable carrier systems. The new L5 carrier system was introduced in 1973. The advent of new and improved solid-state techniques, including integrated circuits, was also instrumental to the development of the L5 system. This analog and FDM-type carrier system has a capacity of 10,800 VF channels per pair of copper tubes. Thus, the largest size 22-tube coaxial cable sheath can handle 108,000 voice channels, using ten working tube pairs with one pair for a protection circuit.

The L5 carrier, as its predecessors the L3 and L4 carriers, is designed to provide high-grade signal-to-noise quality circuits, for up to 4,000 miles (6,437 km). The 10,800-VF channels transmitted over each copper tube pair utilize the 3,124-60,556-kHz spectrum. The attenuation of these copper tubes varies greatly with frequency and significantly with temperature. Therefore, it is necessary to install repeaters at approximately 1-mile (1.61 km) intervals to obtain the required signal-to-noise performance. The dc power for the repeaters is fed over the center conductor in each coaxial copper tube from both ends of a power-feed point. To minimize the line voltage requirements, the maximum length between power-feed points is held to 75 miles (120.7 km). This distance is about one-half of the power-feed span used for the L4 system. Thus, the maximum line voltage is maintained at 1,150 V for a current flow of about 910 milliAmperes (mA) to power approximately 74 remote repeaters located in manholes.

The frequency allocation plan of the L5 system is outlined in the drawing. There are three *jumbogroups,* each comprising 3,600 VF channels, for a total of 10,800 voice channels. Every *jumbogroup* consists of six *mastergroups,* as described for the L4 carrier system. Each basic *jumbogroup* in the L5 occupies the 564-17,548-kHz band. To achieve the L5 frequency spectrum, *jumbogroup* No. 1 is first modulated by a 42,496-kHz carrier. This carrier and the upper sideband products of modulation are suppressed. The resultant lower sidebands, 24,948-41,932 kHz, are next modulated by a 45,056-kHz carrier. The consequent lower sidebands, 3,124-20,108 kHz, form the line frequency band for *jumbogroup* No. 1.

The second *jumbogroup* is first modulated by a 70,144-kHz carrier. The lower sideband products of this modulation step, 52,596-69,580 kHz, are further modulated by the 91,648-kHz carrier frequency. The resultant lower sideband products of this second modulation stage, 22,068-39,052 kHz, become the line frequency of *jumbogroup* No. 2. The third *jumbogroup,* however, is multiplexed by three successive modulation steps, employing carrier frequencies 42,496, 45,056, and 40,448 kHz, respectively, as depicted in the drawing. The lower sideband products of the 42,496-kHz modulator stage are multiplexed by the following second stage's carrier of 45,056 kHz. Its lower sideband products, 3,124-20,108 kHz, are next modulated by the third

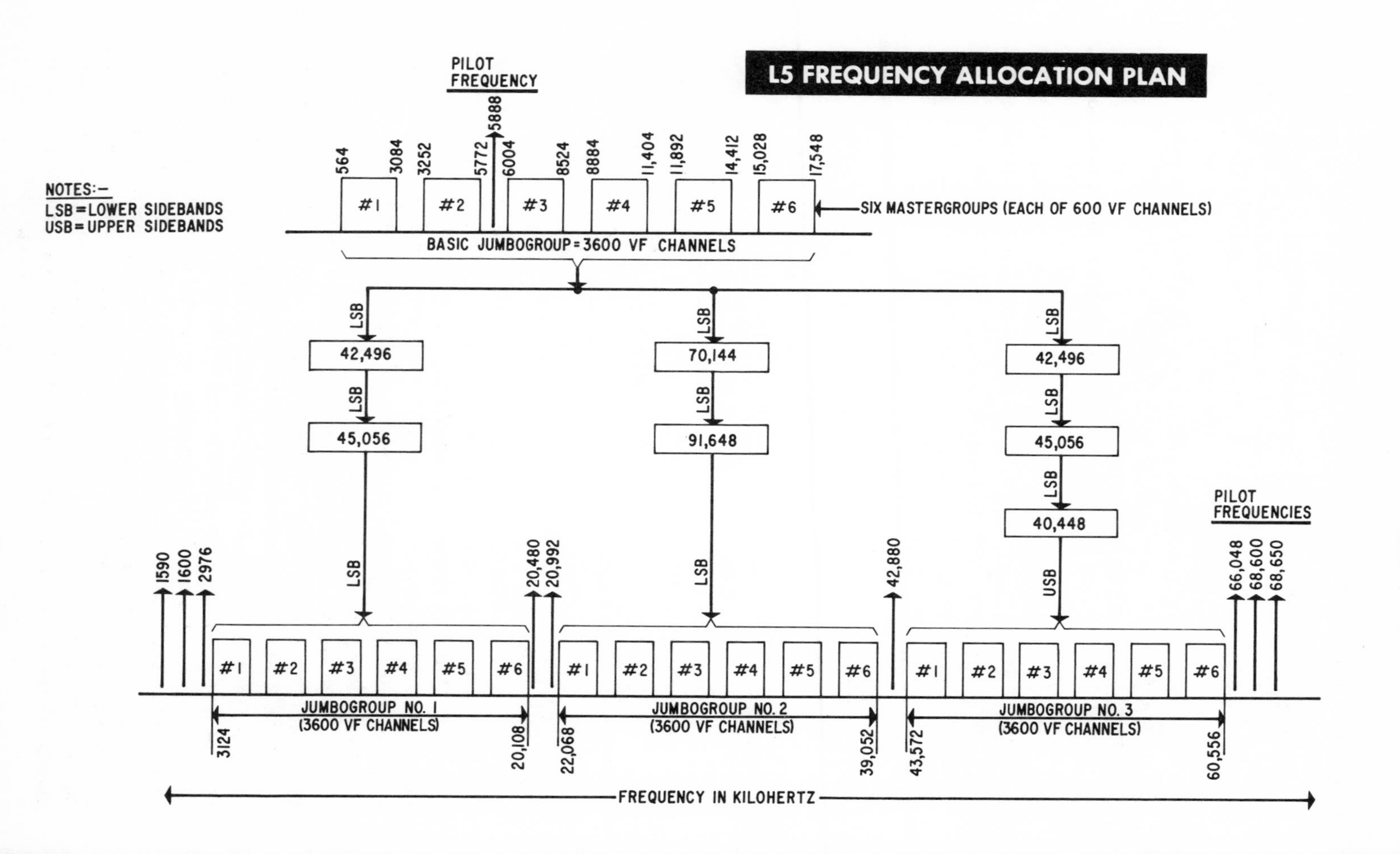
L5 FREQUENCY ALLOCATION PLAN
PILOT FREQUENCY
5888
564
3084
3252
5772
6004
8524
8884
11,404
11,892
14,412
15,028
17,548
#1
#2
#3
#4
#5
#6
SIX MASTERGROUPS (EACH OF 600 VF CHANNELS)
NOTES:—
LSB = LOWER SIDEBANDS
USB = UPPER SIDEBANDS
BASIC JUMBOGROUP = 3600 VF CHANNELS
LSB
42,496
45,056
70,144
91,648
40,448
USB
PILOT FREQUENCIES
1590
1600
2976
20,480
20,992
42,880
66,048
68,600
68,650
JUMBOGROUP NO. 1 (3600 VF CHANNELS)
JUMBOGROUP NO. 2 (3600 VF CHANNELS)
JUMBOGROUP NO. 3 (3600 VF CHANNELS)
3124
20,108
22,068
39,052
43,572
60,556
FREQUENCY IN KILOHERTZ

Coaxial Cable Carrier Systems (cont.)

modulation stage, employing a carrier of 40,448 kHz. Note that the *lower* sidebands and carrier in this stage are suppressed. The *upper* sideband products of 43,572-60,556 kHz are employed for the line frequency of *jumbogroup* No. 3.

The frequency band occupied by the three *jumbogroups*, comprising the 10,800 voice channels of the L5 carrier, is therefore 3,124 to 60,556 kHz. In addition, a number of pilot frequencies are used for various functions, such as fault-locating signals (1,590, 1,600, 68,600, and 68,650 kHz, respectively) and control of repeater equalizing operations (2,976 and 66,048 kHz). The L5 system can also handle high-speed digital data traffic with appropriate terminal equipment, but one *supergroup* of 600 VF channels is displaced for this purpose.

A summary of the main characteristics of the L1 to L5 coaxial cable carrier systems appears in the table.

		VF Channels			
Type	*In service*	*Per copper tube*	*Per coaxial cable sheath*	*Repeater spacing miles (km)*	*Electronic equipment*
L1	1941	600	1,800	8 (12.8)	Vacuum Tubes
L3	1953	1,860	9,300	4 (6.4)	Vacuum Tubes
L4	1967	3,600	32,400	2 (3.2)	Transistors
L5	1973	10,800	108,000	1 (1.61)	Integrated Circuits

Carrier Alarms and Protection Systems

Any failure of the transmission medium or in the common equipment of a carrier terminal must immediately activate protection circuits and alarm systems, thus alerting maintenance personnel. Such actions are imperative because of the great number of voice, data, and video channels that could be deleted. These safeguard requirements are particularly applicable to multiconductor cables, coaxial cables, and microwave radio transmissions. For instance, in cable carrier systems, a break in the conductivity of the cable conductors or an appreciable loss in the received multiplex signals would operate alarm and related protection circuits at associated carrier terminals. Likewise, the loss or a substantial attenuation of the received signals in a microwave radio system would initiate similar protection and alarm operations at microwave terminals and repeater stations.

To provide service continuity during an equipment outage or other disruption, the carrier alarm system is arranged to make-busy the specific trunks connected to the defective carrier channel bank or group. Moreover, the trunks in each toll group are connected to carrier

Carrier Alarms and Protection Systems (cont.)

channels in different channel groups and supergroups. In this manner, greater reliability of service is achieved because the failure of one channel group or supergroup will only affect a portion of the trunks in a particular trunk group. Also, additional service reliability is attained by redundancy of equipment. For example, in the GTE Lenkurt 46A3 carrier system, common equipment items such as group, supergroup, mastergroup units, and all carrier and pilot generators are usually duplicated. Similar redundancy of common equipment is provided in Western Electric L3, L4, and L5 carrier systems.

In broadband carrier systems, separate cable pairs or coaxial copper tubes are normally utilized as a protection circuit. The activation of the carrier alarm system also initiates the transfer of the carrier transmissions from the defective common equipment units or paths to the protection facilities. In this way the faulty equipment or deficient transmission path will be by-passed. When the trouble has been cleared, the protection circuit releases and carrier transmissions proceed in the normal manner.

Carrier Requirements for Radio Transmission

Carrier equipment for use with microwave or other point-to-point radio systems, including earth satellite stations, is similar in many respects to the equivalent cable carrier type. Certain components and functions of wire carrier systems, however, are not needed for radio transmission. For example, the radio transmitting and receiving equipment give the necessary power output and amplification, respectively, to overcome the transmission path attenuation.

For this reason, amplifier regulation in the carrier equipment at

CARRIER SYSTEM USED WITH MICROWAVE RADIO CIRCUIT

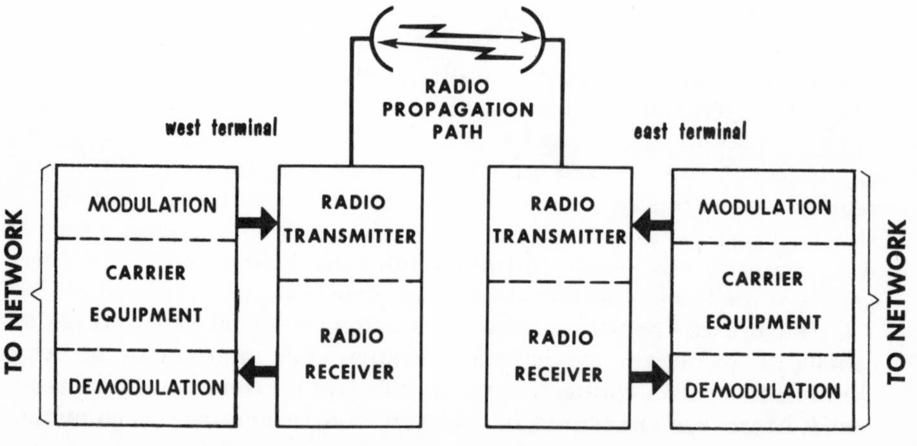

RADIO
PROPAGATION
PATH
west terminal
east terminal
TO NETWORK
MODULATION
CARRIER
EQUIPMENT
DEMODULATION
RADIO
TRANSMITTER
RADIO
RECEIVER
RADIO
TRANSMITTER
RADIO
RECEIVER
MODULATION
CARRIER
EQUIPMENT
DEMODULATION
TO NETWORK

Carrier Requirements for Radio Transmission (cont.)

terminals and repeater points is not required. The radio equipment is usually designed to overcome the noise caused by fading or variations in the radio propagation path so that compandors are not normally provided for carrier systems used with radio circuits. It is also not necessary to include in the carrier equipment the components and circuitry for slope and temperature compensation as with cable carrier systems.

Microwave radio systems in the Bell System's intertoll or Direct-Distance-Dialing (DDD) network carry a large percentage of toll and long distance traffic. Most of these point-to-point microwave systems have a capacity of 6 to 11 radio channels per path or route, depending upon the type of radio equipment and the radio-frequency bands employed. As many as 1,860 two-way voice circuits can be transmitted by one microwave channel. Thus, it is possible to convey approximately 11,160-19,800 voice channels along a particular microwave route. To assure continuity of service, one or more microwave protection channels are provided to parallel the working channels. Therefore, in the event of a transmission outage or equipment failure in any part of a route, a protection channel will immediately take over the traffic from the abortive microwave channel.

The terminals of these microwave radio systems are interconnected with the DDD network by broadband multiplex equipment, such as described for the L3, L4, and L5 coaxial cable carrier systems. For instance, the frequency band of the three mastergroups and one supergroup which embrace the 1860-VF channels in the L3 carrier is 312-8,284 kHz. This frequency range, consequently, would be used to modulate the carrier of the radio transmitter in the microwave terminal. At the distant receiving microwave terminal, the radio signals are first demodulated to the 312-8,284-kHz line frequency of the L3 carrier system. This recovered line frequency band is then applied to the L3 multiplex equipment for demodulation into voice-frequencies. The block diagram illustrates a simple form of these operations. A similar process is employed with L4 multiplex equipment except that only its first threee mastergroups, comprising 1,800-VF channels, would normally be utilized. These three mastergroups occupy the 564-8,524-kHz spectrum and, therefore, only this frequency band would be used to modulate the microwave radio equipment.

Radio Path Losses

Before we look at typical carrier systems used with microwave radio circuits, we should understand something about the characteristics of the transmission path. Attenuation of radio transmissions in *free space*, particularly at microwave frequencies, increases uniformly with distance. Attenuation or path loss also increases with higher radio frequencies. The *free space* condition is considered

Radio Path Losses (cont.)

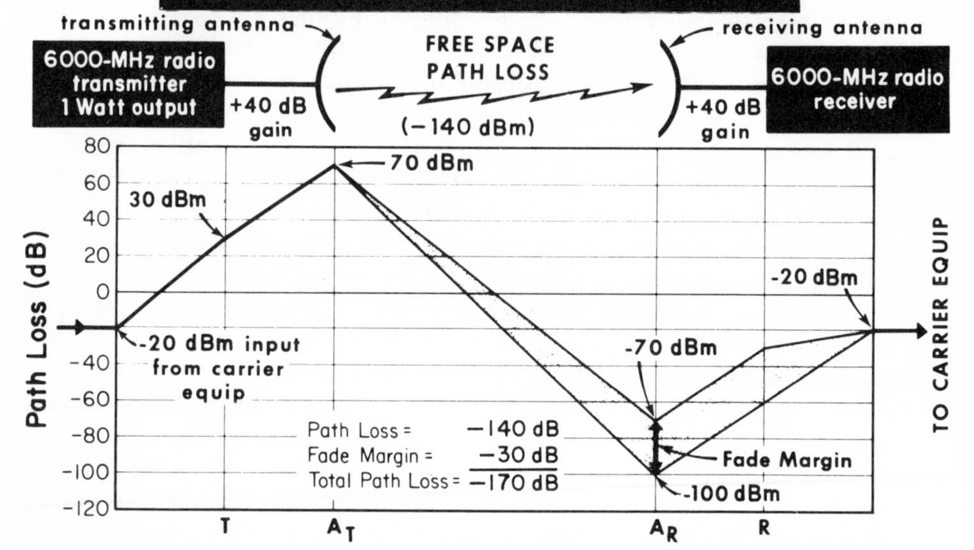

to exist where there is direct line of sight between the transmitting and receiving antennas. This also assumes that the intervening terrain is smooth and free of obstacles to the radio path. This is more theoretical than a practical concept. It is known that atmospheric conditions and the type of terrain can cause considerable fading or variations in the radio transmission path, especially at microwave frequencies. Most telephone company microwave circuits operate in the 4,000-, 6,000-, and 11,000-MHz common carrier bands assigned by the FCC. The path losses at these frequencies are not constant. The radio equipment must thus be able to provide sufficient gain to overcome anticipated variations in the received signal levels.

The power level diagram is of a typical microwave radio circuit in the 6,000-MHz band. It may be compared to the power level diagram for cable carrier systems (page 139). Note that variations in the cable attenuation were compensated for automatically by the carrier equipment. For microwave circuits, the radio equipment is designed to compensate for attenuation changes, up to a predetermined design level, in the transmission path.

Microwave Radio Facilities

The design characteristics of a microwave radio system, including the radio-frequency band used, determine its broadband capabilities and voice channel capacity. Frequency modulation (FM) is the

Microwave Radio Facilities (cont.)

type of emission normally employed. In order to achieve the improved signal-to-noise feature of frequency modulation, the output radio-frequency bandwidth should be more than twice that of the highest carrier frequency in the baseband. For example, the Western Electric type TH-1 microwave radio system has a capacity of 1,860 two-way VF channels. Its associated carrier equipment, equivalent to the L3 type, has a range of 312-8,284 kHz. Thus, a radio-frequency bandwidth of almost 30 MHz is required in the assigned 5,925-6,425-MHz common carrier band for each direction of transmission. Radio-frequency channels of considerable bandwidth, therefore, are needed to handle multichannel carrier systems, television circuits, high-speed data, or other broadband communications.

During the past years the rapid growth of point-to-point microwave systems, including the requirements for satellite communications, has about exhausted the radio-frequency allocations available for common carrier service. Recent improvements in single sideband (SSB) modulation techniques and equipment have made it feasible to utilize SSB modulation instead of FM for microwave radio systems. Since the radio-frequency bandwidth requirements of SSB are substantially less than for the present FM microwave systems, it is anticipated that the present VF channel capacities of microwave systems (see table) can be increased almost four times. This can be accomplished without additional radio-frequency channels, thereby conserving the radio-frequency spectrum for future requirements.

Microwave Radio Facilities Used by Common Carriers

Microwave Bands Assigned by FCC	*Authorized Bandwidth for Each Radio Channel*	*VF Channel Capacity per Radio Channel*
3,700-4,200 MHz	20 MHz	1,500
5,925-6,425 MHz	30 MHz	1,860
10,700-11,700 MHz	50 MHz	2,400

Basic Carrier Equipment for Microwave Radio Channels

The number of VF channels that can be multiplexed in a carrier system for transmission over a microwave radio system depends upon the modulation bandwidth or baseband frequency range of the radio equipment. For instance, the GTE Lenkurt 46A3 and the Western Electric type U600 multiplex equipments have basebands of 564-3,084 kHz for their capacities of 600 VF channels, or one mastergroup. For a baseband of 312-8,524 kHz as in the L3 carrier system, up to three mastergroups may be multiplexed to handle 1,860 VF channels. This baseband range necessitates the use of microwave systems operating in the present 4,000-, 6,000-, or 11,000-MHz frequency bands. The

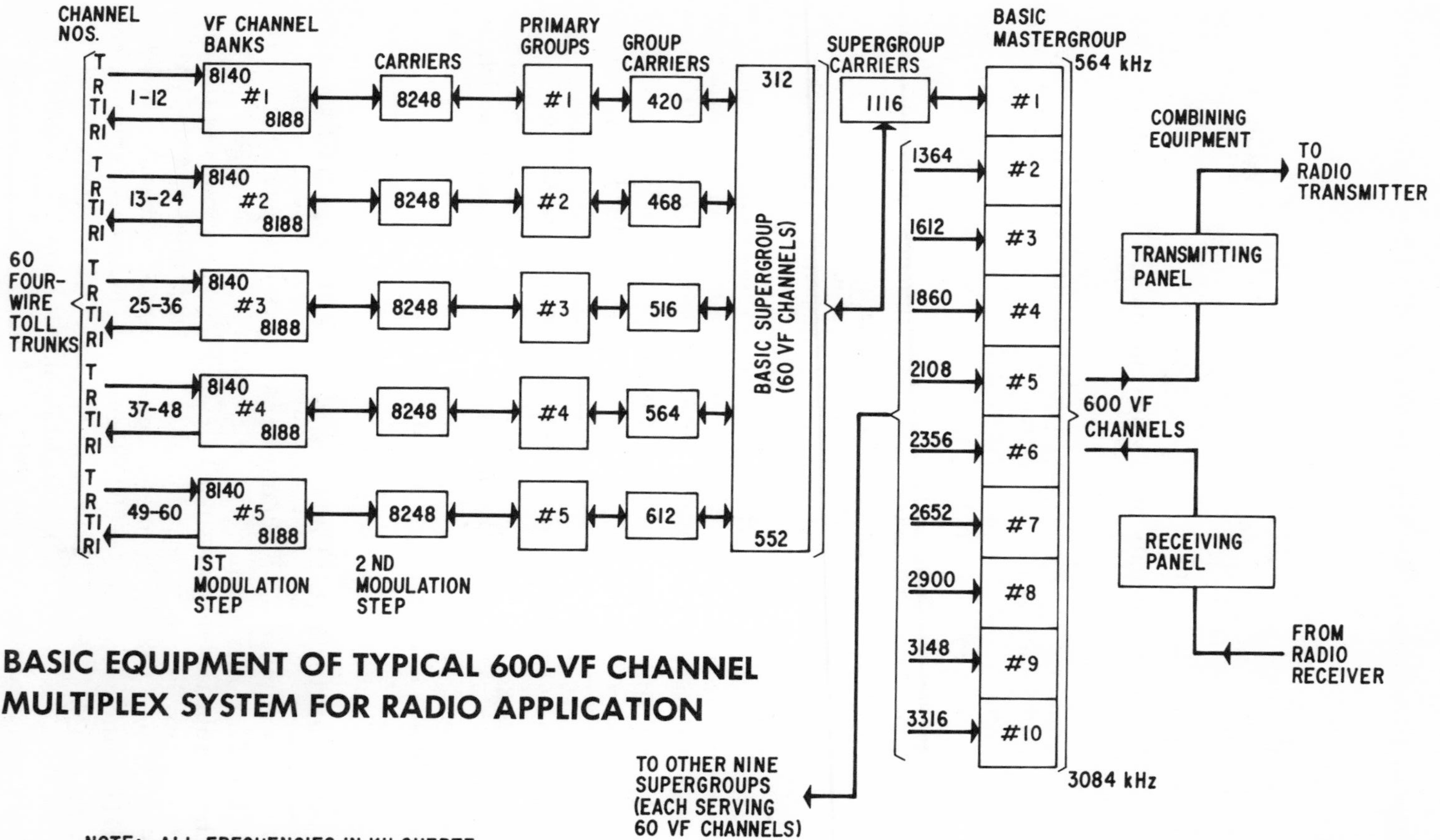

BASIC EQUIPMENT OF TYPICAL 600-VF CHANNEL MULTIPLEX SYSTEM FOR RADIO APPLICATION

Carrier Equipment for Microwave Radio Channels (cont.)

FCC has allocated radio channels in these bands that have the required bandwidths for broadband carrier systems.

Radio-frequency channels in the 2,000-MHz band, which have smaller authorized bandwidths, may be utilized for light-route microwave systems of less than about 100 VF channels. For example, the GTE Lenkurt 36A multiplex equipment has a line frequency range of 12-300 kHz for its capacity of 72 voice channels. This multiplex unit also may be combined with the GTE Lenkurt 46A carrier or Western Electric U600 multiplex equipment to furnish one or more supergroups (a supergroup has a capacity of 60 VF channels), if more than 72 voice circuits are to be handled. Likewise, ten supergroups may be combined to form one mastergroup, as shown in the diagram on the facing page: This typical multiplex system, like the GTE Lenkurt 46A3 or Western Electric U600, may be used to interconnect 600 VF channels with a microwave radio system. Its operation is described later. These multiplex systems are also of the single-sideband suppressed carrier type which utilize solid-state electronic devices. They employ modulation plans similar to those described for the L3 and L4 coaxial cable systems.

Transmitting Operations of Channel Unit

Recall that in the section on FDM Cable Carrier Systems, the basic equipment and operation of the older type GTE Lenkurt 45BN carrier system were described to aid the reader in understanding cable carrier systems. A similar procedure will be followed in tracing the paths of VF and related carrier signals through the channel units and other components of the GTE Lenkurt 36A and 46A3 multiplex equipments. The operations are similar to those of the Western Electric U600 and related multiplex equipment.

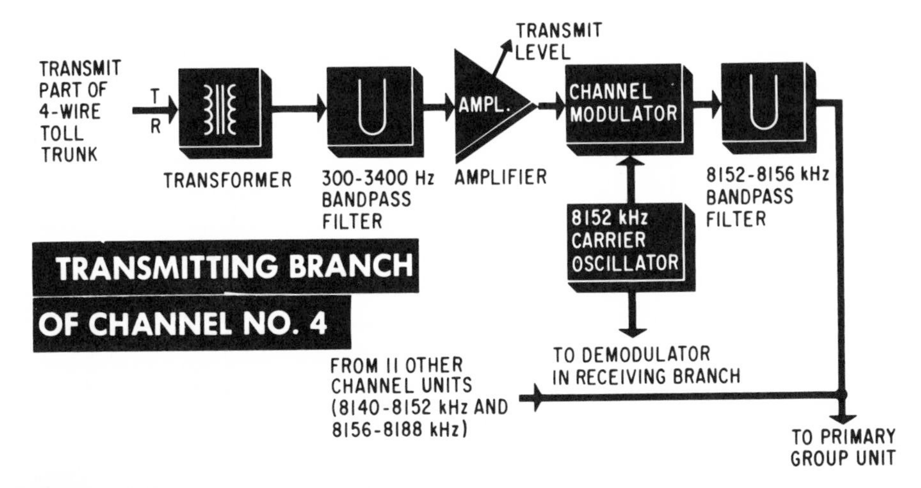

Transmitting Operations of Channel Unit (cont.)

Referring to the representative schematic diagram on p. 195, VF signals from a four-wire toll trunk enter the assigned channel unit in a particular 12-channel bank, through a balanced 600-ohm transformer. The signals may first go through a bandpass filter, or directly to a *limiting amplifier*. By adjusting this amplifier, each channel's output signal is subsequently set to the same level. The bandpass filter, which usually consists of a 300-Hz high-pass filter and a 3,400-Hz low-pass filter in tandem, is used to preclude interference from low frequency and dc switching transients. The resultant voice signals in the 300-3,400-Hz range are then amplified before being applied to the channel modulator for up-conversion.

The voice signals are modulated by one of 12-channel carriers, each spaced 4 kHz in the 8,140-8,184-kHz range, depending upon its particular channel number. For example, channel No. 1 is modulated by the 8,140-kHz carrier, channel No. 2 by 8,144 kHz, channel No. 3 by 8,148 kHz, etc. to 8,184-kHz carrier for channel No. 12. Let us assume that our VF signal is in channel No. 4. The 8,152-kHz carrier, therefore, will be modulated. The modulation products will pass through the *transmit bandpass filter*. Only the upper sideband (8,152.3-8,155.4 kHz) will be selected for transmission to the *primary group* unit which is common to channels 1 to 12 inclusive. The lower sideband and other undesired products of the modulation process are attenuated by the bandpass filter.

The channel carrier frequencies noted above are initially generated in the *primary group* unit. The 8,248-kHz frequency of the *primary group carrier* is produced from a crystal-controlled oscillator. It is then mixed with a generated pulse spectrum in the 64-108-kHz range. The lower sideband products, at 4-kHz steps in the 8,140-8,184-kHz band, are applied to a crystal filter in each channel. The required modulation and demodulation frequencies are selected in this manner for the 12 channels in each channel bank.

Transmitting Operations of Primary Group Unit

Twelve channel banks or units comprise a *primary group*. Channel No. 4 thus will be in *primary group* No. 1, together with the other 11 channels in the Channel 1–12 series. Each of these channels occupies a 4-kHz segment in the 8,140-8,188-kHz band, as previously noted. The second modulation step takes place in the primary group units. The 8,140-8,188-kHz band of carrier signals from channels 1-12 is modulated by an 8,248-kHz carrier, as shown on the next page, top.

The modulation products are 60-108 kHz and 16,388-16,436 kHz for lower and upper sidebands, respectively. The lower sideband (60-108 kHz) is selected by the low-pass filter following the modulated stage; the upper sideband is attenuated. The lower sideband is then

Transmitting Operations of Primary Group Unit (cont.)

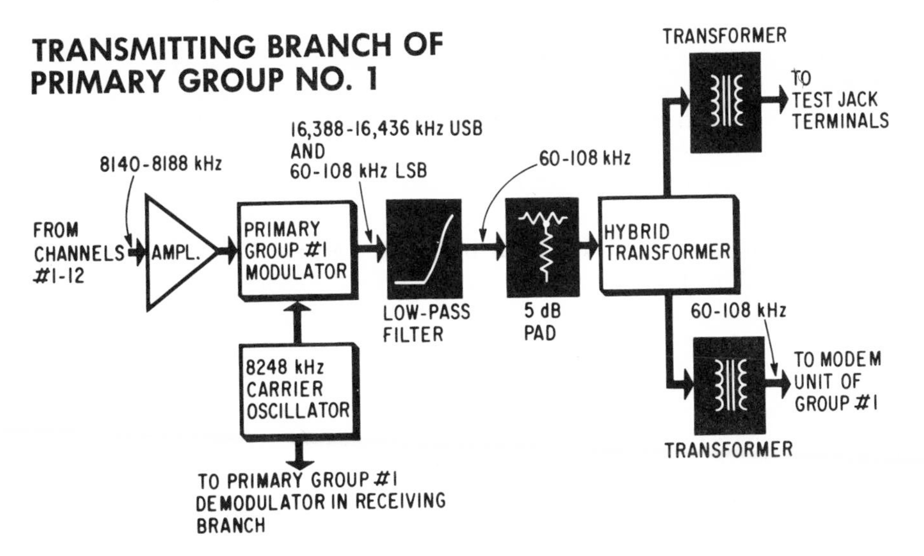

passed through a 3.5 dB resistor pad to an output *hybrid* transformer. This type of transformer is symmetrical in design. Thus, it applies equal 60-108-kHz carrier power to the associated *group unit* and to the transmitting test jack terminals.

The other four series of 12-channel units will function in a similar fashion within their respective primary group units. Channels 13-24 are associated with group 2, channels 25-36 with group 3, channels 37-48 with group 4, and channels 49-60 with group 5. These assignments are also shown in the previous block diagram of a typical 600-VF channel multiplex system (page 194).

Functions of Transmission Branch of Group Modem Unit

The 12 channels of primary group No. 1 were multiplexed or translated to the 60-108-kHz band by the two modulation steps of this primary group, as previously described. These carrier signals are next directed to the *transmitting branch* of the *group modem unit*. They enter the modem unit through the input transformer and pass through a low-pass filter. This filter serves to attenuate frequencies in the radio broadcast band, as well as other frequencies above 108 kHz that may be present.

The carrier signals then may pass through a 5-dB resistor pad or go directly to the 3-dB $\pm$ 1 dB adjustable *transmit level pad*. The 5-dB pad is strapped out when the carrier signal is below a predetermined level, usually $-$ 42 dBm. The transmit level pad permits the group's output signal to be subsequently adjusted to the required level at its related *supergroup's* transmit terminals.

Following this, the 60-108-kHz signals enter the *group* No. 1

Transmission Branch Functions in Group Modem Unit (cont.)

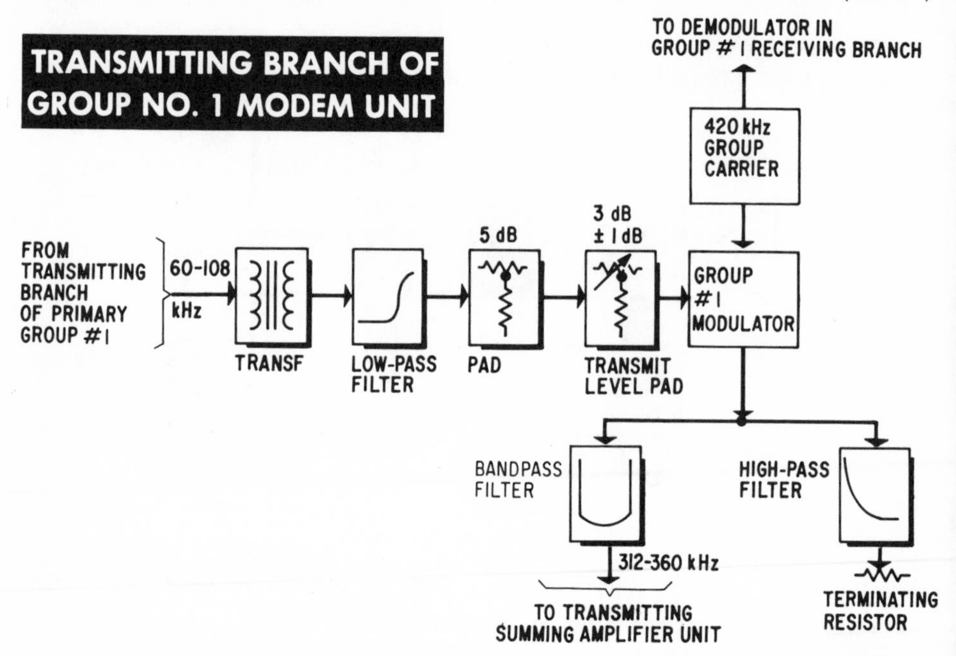

modulator and modulate the 420-kHz *group carrier*. This modulation process produces upper sideband frequencies of 480-528 kHz and lower sideband frequencies of 312-360 kHz. The bandpass filter following the modulator selects the 312-360 kHz lower sideband product. The high-pass filter, which is also connected to the modulator's output, attenuates the unwanted upper sideband and provides the proper impedance termination for efficient transmission of carrier signals. The modem units of the other four groups function in an equivalent manner. Their group carrier modulation frequencies are 468, 516, 564 and 612 Hz respectively, for groups numbered 2, 3, 4, and 5.

Operation of Group Transmit Summing Amplifier Unit

The output of the bandpass filter in each *group modem unit* is connected through a 75-ohm resistor to a common point, designated the *summing bus*. This bus connects to the zero impedance input of the *transmit summing amplifier*. This arrangement serves to eliminate crosstalk or other interactions between the five connecting groups. The *basic supergroup* of 312-552 kHz is formed at the *summing bus* interconnections. This carrier band is next conveyed to the input of the *transmit summing amplifier*, as shown on the next page.

The nominal gain of the transmit summing amplifier is about 28 dB. Its output terminates in a circuit consisting of a 10-dB pad, a 3.5-

Group Transmit Summing Amplifier Unit (cont.)

dB pad, and a 75-ohm impedance output transformer. The 10-dB pad is strapped out if the *supergroup's* transmit level requirement is – 25 dBm. The output transformer provides the termination for the 312-552-kHz *basic supergroup* of 60 VF channels. From two to ten supergroups may be combined by means of additional modulation stages. In this manner, one *mastergroup* comprising ten supergroups may be formed to handle 600-VF channels, as later explained.

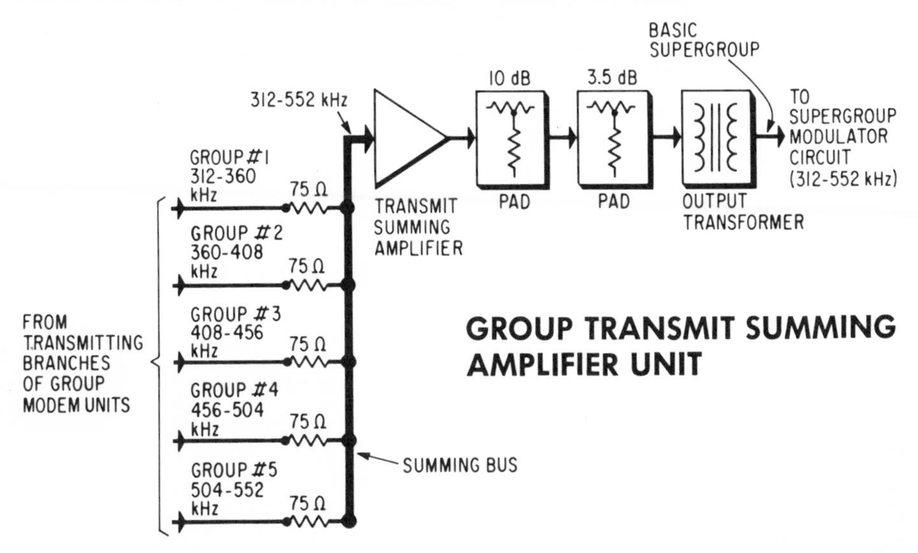

GROUP TRANSMIT SUMMING AMPLIFIER UNIT

Functions of Supergroup Modulator Units

The basic supergroup (312-552 kHz), as previously explained, has a capacity of 60 VF channels. Ten supergroups may be combined to form a *mastergroup,* which will have a capacity of 600 voice channels, to provide additional channel requirements. Likewise, three *mastergroups* may be combined, as in the case of the L3 carrier system, to furnish up to 1,860 VF channels. This is the nominal capacity of the average long-haul microwave radio channel. The combining of a supergroup's signals to form a basic *mastergroup* is the main function of each *supergroup modulator unit.*

Ten *supergroup modulator units* are needed, one for each supergroup, to form the *mastergroup.* Let us follow the path of a 312-552-kHz signal from a 60-channel supergroup (see block diagram on the next page). The signal enters the supergroup modulator unit through a 10-dB pad. This pad is strapped out if the signal level is much below – 25 dBm. The signal passes through a 552-kHz low-pass filter and the *transmit level adjust pad* to the *supergroup modulator.* The low-pass filter attenuates the unwanted spurious signals above 552 kHz. The

Functions of Supergroup Modulator Units (cont.)

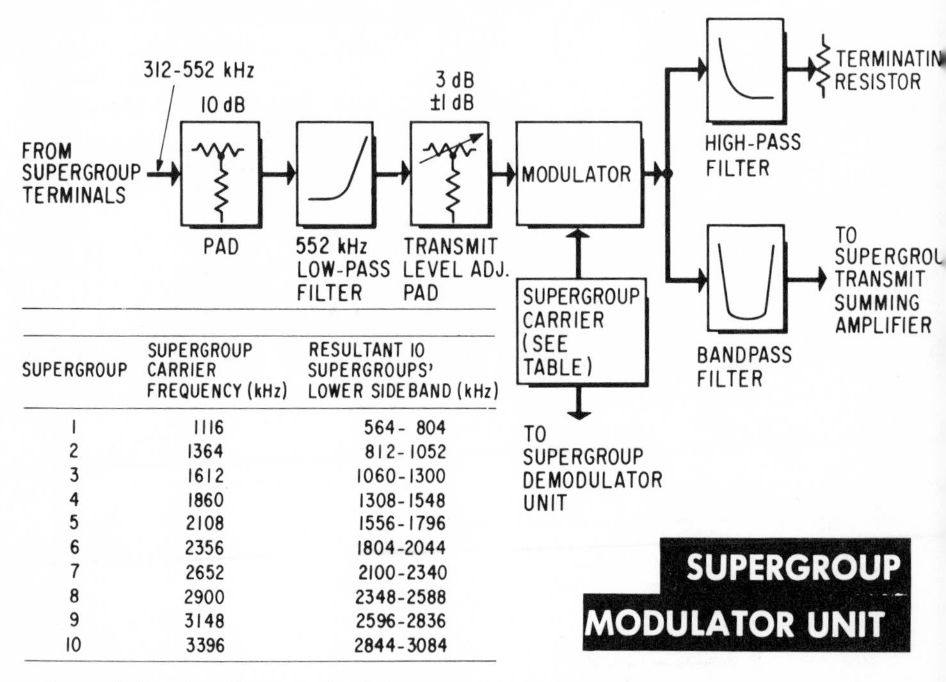

SUPERGROUP	SUPERGROUP CARRIER FREQUENCY (kHz)	RESULTANT 10 SUPERGROUPS' LOWER SIDEBAND (kHz)
1	1116	564- 804
2	1364	812-1052
3	1612	1060-1300
4	1860	1308-1548
5	2108	1556-1796
6	2356	1804-2044
7	2652	2100-2340
8	2900	2348-2588
9	3148	2596-2836
10	3396	2844-3084

transmit level adjust pad adjusts the signal level from the supergroup to compensate for transmission loss and equipment tolerance.

In the *supergroup modulator,* the 312-552-kHz signal from the supergroup is multiplexed or translated by the assigned *supergroup carrier* frequency. Table (A) in the drawing (opposite) lists the ten supergroup carrier frequencies and their lower sideband products. For instance, the assigned supergroup carrier frequency for supergroup No. 4 is 1,860 kHz. The resultant 1,308-1,548-kHz lower sideband will be selected by the bandpass filter which follows the modulator. The high-pass filter, also connected to the modulator's output, provides a proper termination for the unused 2,172-2,412-kHz upper sideband product. This arrangement also improves the *return loss* or transmission efficiency of the circuit. The selected 1,308-1,548-kHz band is now transmitted to the *supergroup transmit summing amplifier circuit.* It will be combined, as shown later, with the designated lower sidebands of the other supergroup modulator units.

Supergroup Combining Operations

The outputs from the ten supergroup modulator units are combined in the *supergroup transmit summing amplifier* equipment. The combining method is diagrammed on the opposite page. Each supergroup's output must first pass through a 75-ohm resistor to a common connection that leads to the input of the *supergroup transmit*

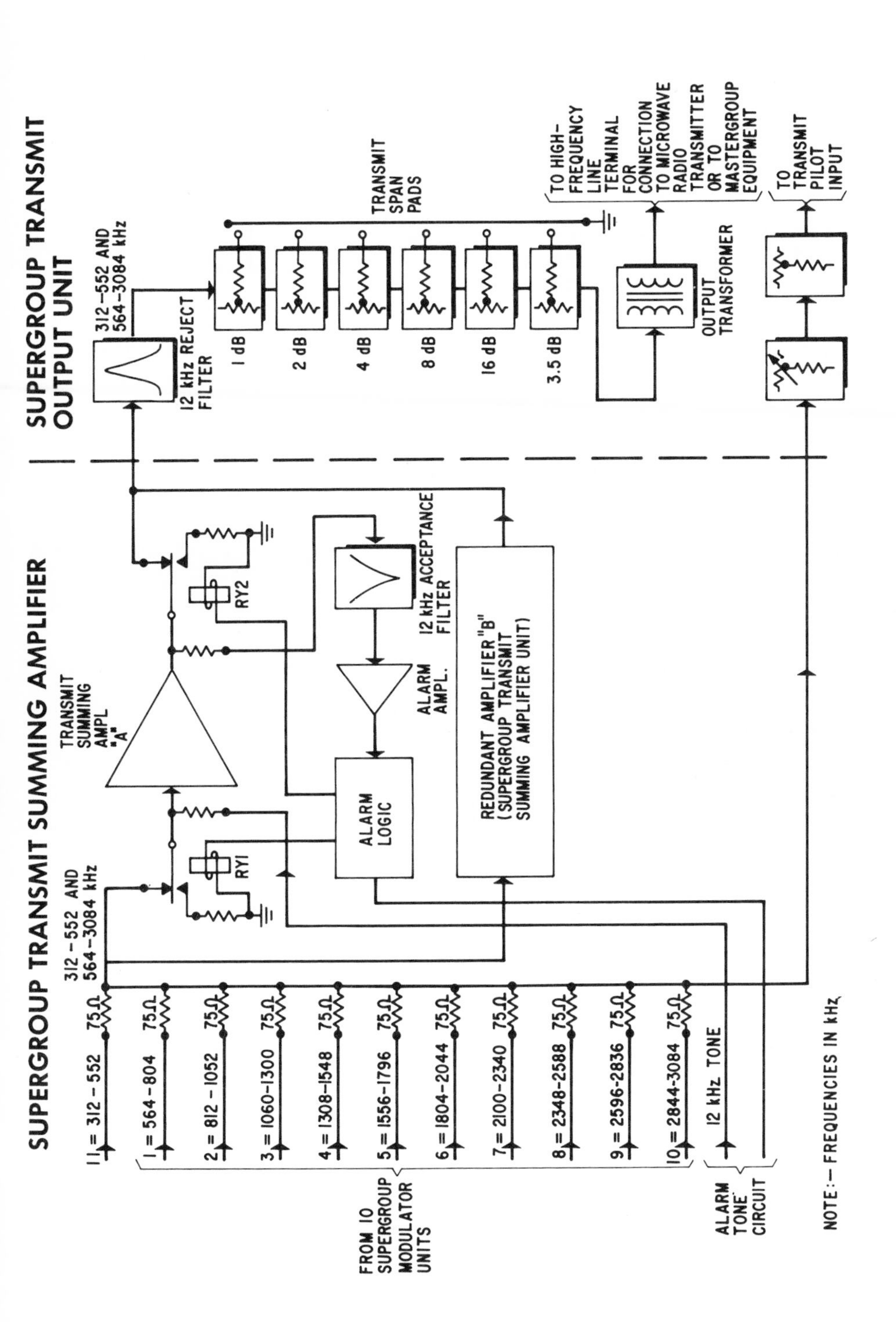
SUPERGROUP TRANSMIT SUMMING AMPLIFIER
SUPERGROUP TRANSMIT OUTPUT UNIT
11 = 312 - 552
1 = 564-804
2 = 812 - 1052
3 = 1060-1300
4 = 1308-1548
5 = 1556-1796
6 = 1804-2044
7 = 2100-2340
8 = 2348-2588
9 = 2596-2836
10 = 2844-3084
75Ω
12 kHz TONE
FROM 10 SUPERGROUP MODULATOR UNITS
ALARM TONE CIRCUIT
312 - 552 AND 564-3084 kHz
TRANSMIT SUMMING AMPL "A"
RY1
RY2
ALARM LOGIC
ALARM AMPL.
12 kHz ACCEPTANCE FILTER
REDUNDANT AMPLIFIER "B" (SUPERGROUP TRANSMIT SUMMING AMPLIFIER UNIT)
12 kHz REJECT FILTER
312-552 AND 564-3084 kHz
1 dB
2 dB
4 dB
8 dB
16 dB
3.5 dB
TRANSMIT SPAN PADS
OUTPUT TRANSFORMER
TO HIGH-FREQUENCY LINE TERMINAL FOR CONNECTION TO MICROWAVE RADIO TRANSMITTER OR TO MASTERGROUP EQUIPMENT
TO TRANSMIT PILOT INPUT
NOTE:— FREQUENCIES IN kHz

Supergroup Combining Operations (cont.)

summing amplifier. This amplifier, as in the case of the group transmit summing amplifier, is designed for a zero impedance input, thus eliminating interactions between the ten connecting supergroups. The gain of the summing amplifier is normally adjusted to provide a nominal output of about 12 dBm.

Because of the large number of voice channels handled (up to 600 VF channels for ten supergroups), duplicate transmit summing amplifier units are often furnished for redundant operation. As shown, a 12-kHz tone from the alarm tone circuit is sent through the usual transmit summing amplifier *A*. This tone is admitted by the 12-kHz band acceptance filter. It is then amplified by the *alarm amplifier* and passed, for monitoring purposes, to the alarm logic circuit. If the level of the 12-kHz tone drops more than 3 dB at any time, the alarm logic circuit operates the two relays, RY1 and RY2. Operation of these relays will immediately switch-out amplifier *A* and switch-in the redundant *B* transmit summing amplifier unit. Finally, audio and visual alarms will also be actuated to alert maintenance personnel.

An additional supergroup (312-552 kHz) may be directly combined with the aforementioned ten supergroups. Thus, a total of 660 VF channels may be connected to the microwave radio terminal. This extra supergroup, designated No. 11 in the drawing, is not modulated in its *supergroup modulator unit,* as will be explained.

From the supergroup transmit summing amplifier, the carrier signals of the combined eleven supergroups are in the 312-3,084-kHz spectrum. They then proceed to the *supergroup transmit output unit.* The signals first pass through the 12-kHz band-reject filter to suppress the 12-kHz alarm tone that is supplied, as previously explained, to the input of the transmit summing amplifier. Next, the carrier signals go through the *transmit span pads* and the *output transformer,* and finally to the high-frequency line transmit terminals. The transmit span pads can be adjusted in 1-dB steps by using plug-in strapping links. This will vary the output level from − 16.5 dBm to − 47.5 dBm, or from − 13 dBm to − 44 dBm when the 3.5-dB pad is strapped out.

Interconnection with Microwave Radio Terminal

The 564-3,084-kHz baseband, containing 600 VF channels, constitutes one mastergroup. This baseband may be combined with a basic supergroup (312-552 kHz) to supply a total of 660 VF channels to a microwave radio terminal. This additional supergroup, as previously indicated, is transmitted to the high-frequency (HF) line terminal without further modulation in its *supergroup modulator unit.* The diagram (opposite page) illustrates the circuit used for this purpose.

The 312-552-kHz signals of supergroup No. 11 pass through the 10-dB pad, the transmit level adjust pad, and the 6.5-dB pad, going

Interconnection with Microwave Radio Terminal (cont.)

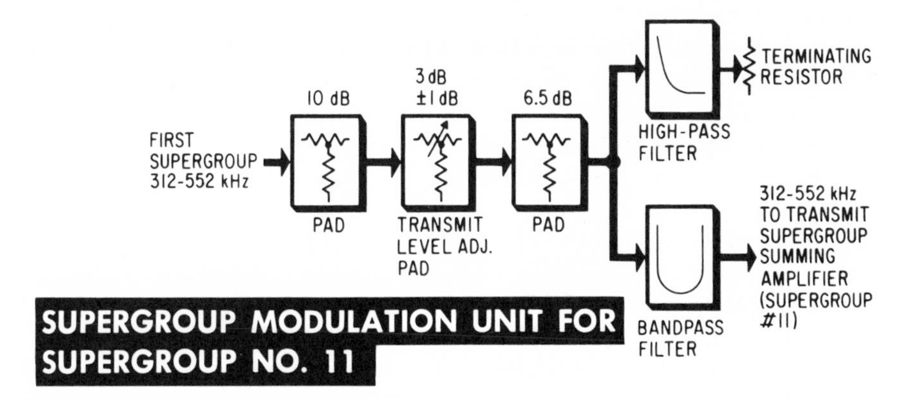

directly to the high-pass and bandpass filters. The carrier signals are not modulated in this supergroup modulator circuit. The high-pass filter suppresses spurious signals above 552 kHz. It is terminated in a resistor that matches its impedance, in order to improve the efficiency of the transmission path. The bandpass filter is designed to only pass the 312-552-kHz band to the following *transmit supergroup summing amplifier.* The transmit level adjust pad is utilized to adjust the signal level of the supergroup to compensate for transmission losses in the circuit. The complete baseband for the 660-VF channels, consequently, will be 312-3,084 kHz, as shown in the preceding block diagram of the supergroup transmit summing amplifier.

Three mastergroups may be combined by employing additional modulation steps, similar to those described for the L3 carrier system. Thus, a total of 1,860 VF channels, including the aforementioned added supergroup No. 11, may be combined for transmission over one microwave radio channel. Interconnection methods with the microwave radio terminal depend upon the distance involved. In the same building or location, standard 75-ohm coaxial cables, one for each direction of transmission, are usually employed. Where the microwave radio equipment is some distance away, it may be necessary to use balanced video pairs in a telephone cable, one for each transmission direction.

Receiving Operations in Broadband Carrier Equipment

The received microwave signals are translated by the microwave radio receiver into the 312-3,084-kHz baseband carrier spectrum. The 660-VF channels, that were originally transmitted, are contained in this baseband. The received baseband signals generally follow a reverse path to that described for the carrier transmitting operations.

The power levels of the received carrier signals are normally in the −8-dBm to −39-dBm range. The received carrier signals first go

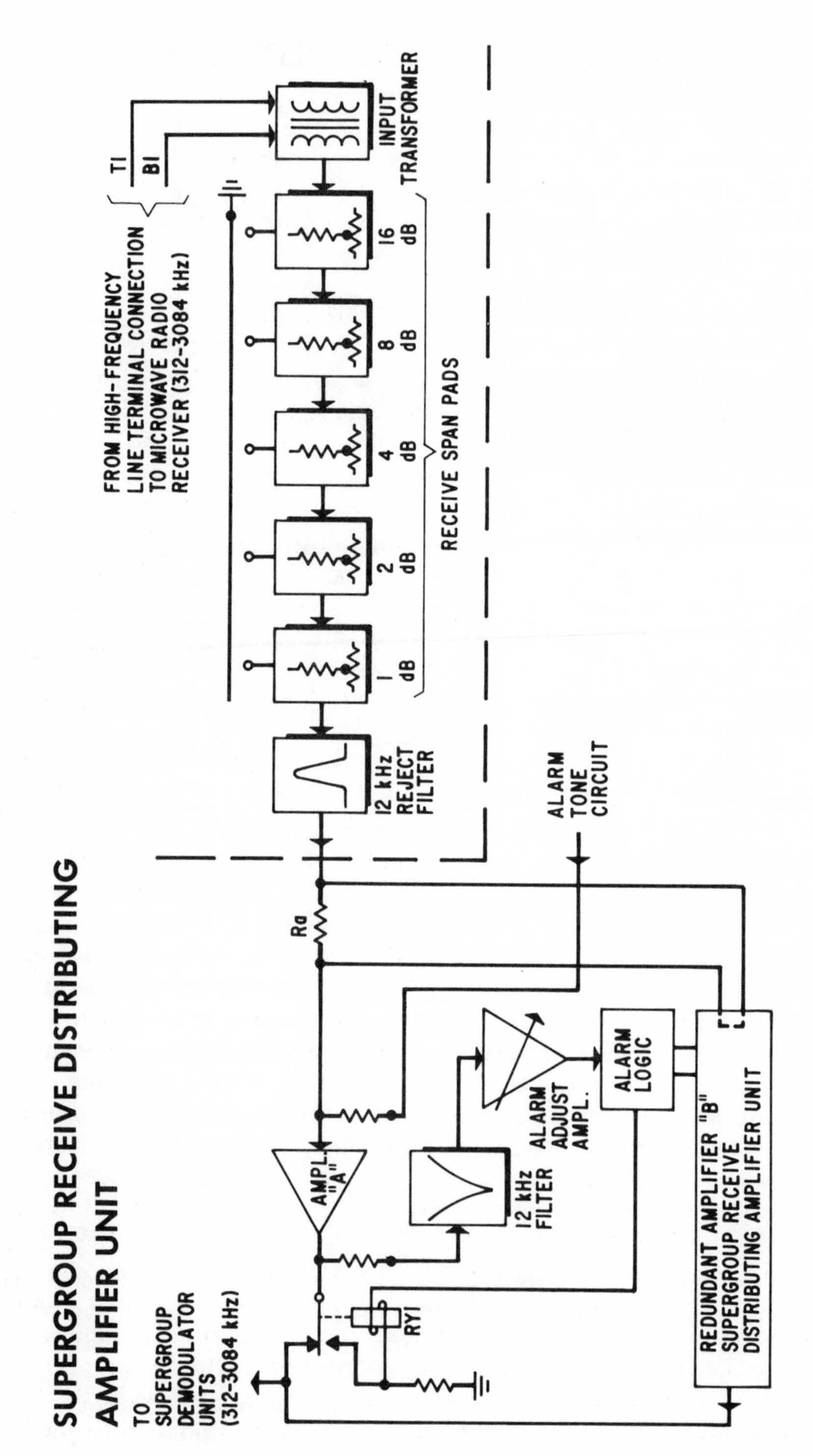
SUPERGROUP RECEIVE DISTRIBUTING
AMPLIFIER UNIT
TO SUPERGROUP DEMODULATOR UNITS (312-3084 kHz)
FROM HIGH-FREQUENCY LINE TERMINAL CONNECTION TO MICROWAVE RADIO RECEIVER (312-3084 kHz)
T1
B1
INPUT TRANSFORMER
16 dB
8 dB
4 dB
2 dB
1 dB
RECEIVE SPAN PADS
12 kHz REJECT FILTER
ALARM TONE CIRCUIT
Ra
AMPL. "A"
12 kHz FILTER
ALARM ADJUST AMPL.
ALARM LOGIC
REDUNDANT AMPLIFIER "B" SUPERGROUP RECEIVE DISTRIBUTING AMPLIFIER UNIT
RY1

Broadband Carrier Equipment Receiving Operations (cont.)

through the input transformer and the *receive span pads* of the *supergroup receive input unit*, as shown on the opposite page. Span pads can provide up to 30 dB of attenuation. Individual resistor span pads may be left in or strapped out, as required, to insure that the input level to the following *supergroup receive distributing amplifier* will be approximately – 39 dBm.

In view of the large number (660) of VF channels handled by the supergroup receive distributing amplifier circuit, two amplifiers are often installed for redundancy operation. The normal path, shown on the schematic, is through amplifier *A*, because series resistor Ra is normally shunted out by the loop circuit in amplifier *B*.

A 12-kHz tone from the alarm tone circuit is sent into amplifier *A* and its associated 12-kHz filter to the alarm logic circuit for monitoring purposes. The 12-kHz *reject filter* prevents this tone from entering the *supergroup receive input unit*. If the 12-kHz tone level from amplifier *A* should at anytime drop by more than 3 dB, the *alarm logic* circuit will be immediately actuated. It will cause the shunt across resistor Ra to be removed, and relay RY-1 will operate. These actions will switch-in amplifier *B* in place of amplifier *A*, as can be seen from the schematic diagram. Simultaneously, maintenance personnel will be alerted by visual and audio alarm signals.

When the trouble condition in amplifier *A* is corrected, the alarm logic circuit will function to release relay RY-1 and to again short-circuit resistor Ra at the input to amplifier *A*. Therefore, amplifier *A* will be switched back into service in place of redundant amplifier B.

Operations of Supergroup Demodulator Unit

From the supergroup receive distributing amplifier unit, the 312-3,084 kHz baseboard carrier signals connect to a common supergroup shelf. This can be equipped with as many as eleven *supergroup demodulator units*. Each unit contains a bandpass filter which is designed to select only that particular 240-kHz band increment of 60 VF channels to which it is tuned. All other frequencies are attenuated. For example, the bandpass filter of supergroup No. 1 selects the 564-804 kHz spectrum, as illustrated on the next page. The frequency ranges of the other ten bandpass filters are listed in the table found below the diagram.

Referring to the drawing, the selected 564-804 kHz carrier signals pass through the related bandpass filter and the adjustable *receive level pad*, and go to the *demodulator*. The carrier frequency applied to the demodulator in each supergroup demodulator unit is such that the resultant lower sideband product always will be in the 312-552-kHz band. For instance, the carrier frequency applied to the de-

Operations of Supergroup Demodulator Unit (cont.)

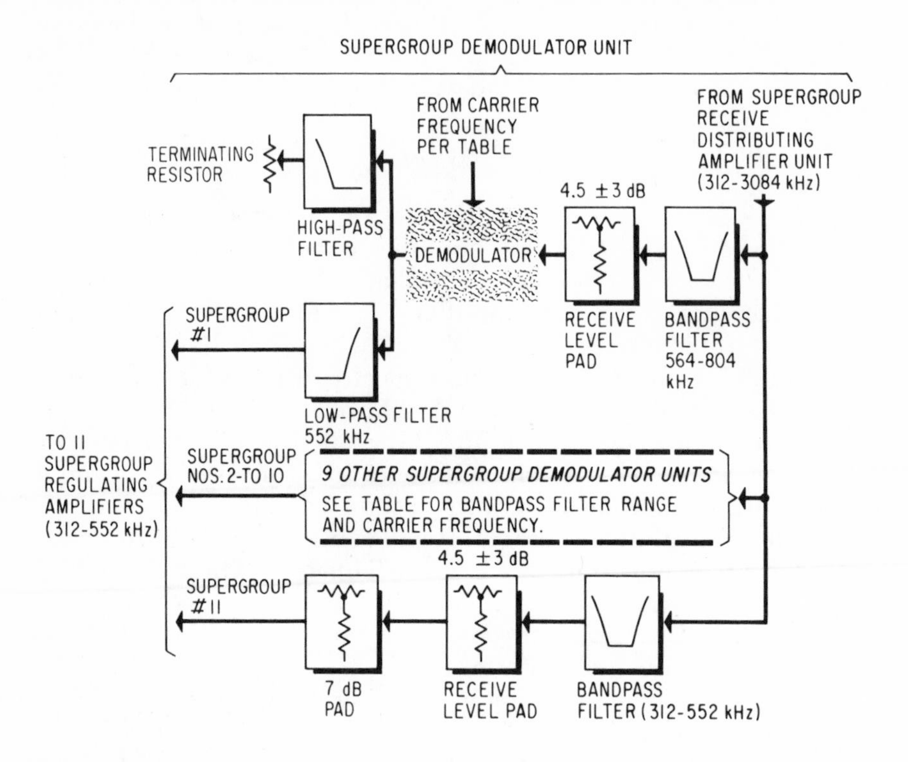

(312-552 kHz) SUPERGROUP NO.	BANDPASS FILTER RANGE	CARRIER FREQUENCY TO DEMODULATOR	(312-552 kHz) SUPERGROUP NO.	BANDPASS FILTER RANGE	CARRIER FREQUENCY TO DEMODULATOR
1	564- 804	1116	7	2100-2340	2652
2	812-1052	1364	8	2348-2588	2900
3	1060-1300	1612	9	2596-2836	3148
4	1308-1548	1860	10	2844-3084	3396
5	1556-1796	2108	11	312- 552	—
6	1804-2044	2356			

modulator in supergroup No. 1 is 1,116 kHz. The other carrier frequencies for supergroups 2 to 10, inclusive, are also listed in the table.

From the demodulator, a low-pass filter passes the 312-552-kHz signals to the output terminals of the supergroup demodulator unit. A high-pass filter is also connected to the output of the demodulator. This serves to terminate the 1,680-1,920-kHz upper sideband product of the demodulation process. Note that demodulation equipment is not provided for supergroup No. 11 because its bandpass filter directly selects the 312-552-kHz spectrum. A 7-dB pad is provided in this particular supergroup unit in place of the demodulator equipment and the low-pass and high-pass filters, as shown in the block diagram.

Functions of Supergroup Receive Regulating Amplifier

Each output of the eleven supergroup demodulator units connects to its own *supergroup receive regulating amplifier.* The prime function of this regulating amplifier is to automatically adjust the gain of the 312-552-kHz carrier signals (containing 60 VF channels) with respect to the incoming supergroup pilot level. The pilot frequency for controlling the operation of the supergroup receive regulating amplifier may be 315.92, 411.91 or 547.92 kHz, as determined by the particular carrier system.

The arrangement of a typical supergroup receive regulating amplifier is shown in the block diagram. The pilot filter, which is of the acceptance type, is designed to admit only the desired 411.92-kHz-pilot frequency from the output of the regulating amplifier. This received pilot frequency is amplified and applied to the *pilot range and alarm circuits.* These circuits detect the pilot frequency and, after rectification, cause the resultant dc voltage to control a *variolosser,* as indicated in the drawing. The variolosser is designed to increase or decrease its attenuation in the path between the two fixed-gain amplifier stages of amplifier *A*.

This regulation action serves to keep the supergroup receive output level within at least ±0.2 dB, with respect to incoming pilot levels that may vary as much as ±6dB from the required value. If the incoming pilot level should drop by more than 7 dB, the variolosser will be shunted out and the regulating amplifier will function as a two-stage fixed-gain amplifier.

The supergroup receive regulating amplifier is also arranged for

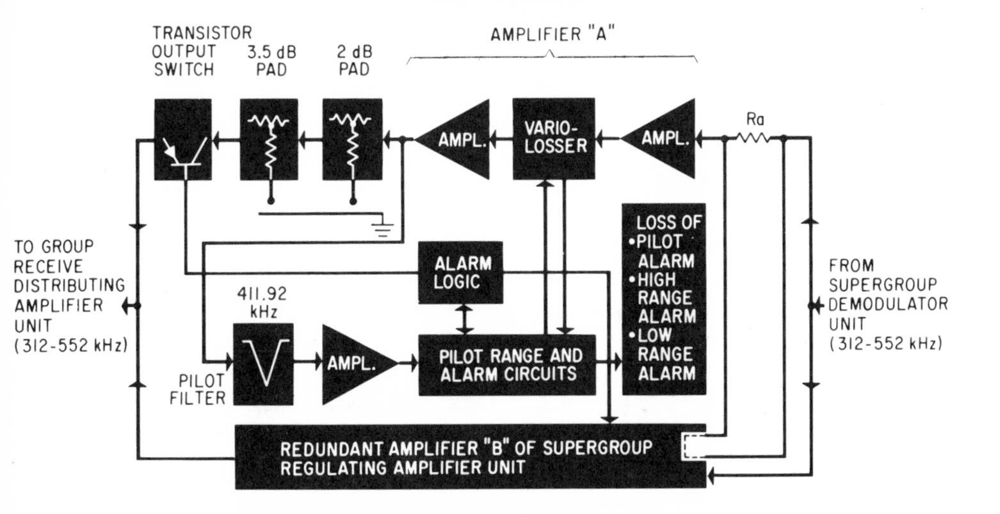

Supergroup Receive Regulating Amplifier Functions (cont.)

redundant operation in a manner similar to that of the supergroup receive distributing amplifier unit. Amplifier *A* normally is in service because, as shown in the diagram, series resistor, Ra, is shunted by a circuit in the input of amplifier B. Furthermore, the transistor output switch connects the output of amplifier *A* to the *group receive distributing amplifier unit.* In the event the pilot level should drop below a pre-set level, the alarm logic circuit will be forced to function. The shunt across resistor Ra will be removed and the transistor output switch will open the path from amplifier *A*. The incoming carrier signals, therefore, will be directed to amplifier *B,* and maintenance personnel will be alerted by visual and audio alarm signals. When amplifier *A* is repaired it will be returned to service by the alarm logic circuit, associated operations in amplifier *B,* and the release of the transistor output switch.

Group Receive Distributing Amplifier Unit

From the supergroup receive regulating amplifier unit, the 312-552-kHz carrier signals enter the *group receive distributing amplifier unit.* Five groups, each containing 12 VF channels, are in every basic supergroup's carrier band of 312-552 kHz. The diagram below shows the components of a group receive distributing amplifier unit.

The 312-552-kHz signals from the supergroup are coupled to the distributing amplifier by a 75-ohm input transformer. They must pass through a 4-dB pad and an optional 2-dB pad. The 4-dB pad isolates the distributing amplifier unit from the supergroup equipment. The 2-dB pad is inserted in the circuit when the supergroup receive level is −28 dBm. This 2-dB pad is strapped out if the level is −30 dBm. The output of the distributing amplifier connects, through 75-ohm resistors, to each of five group modem units, as depicted in the schematic

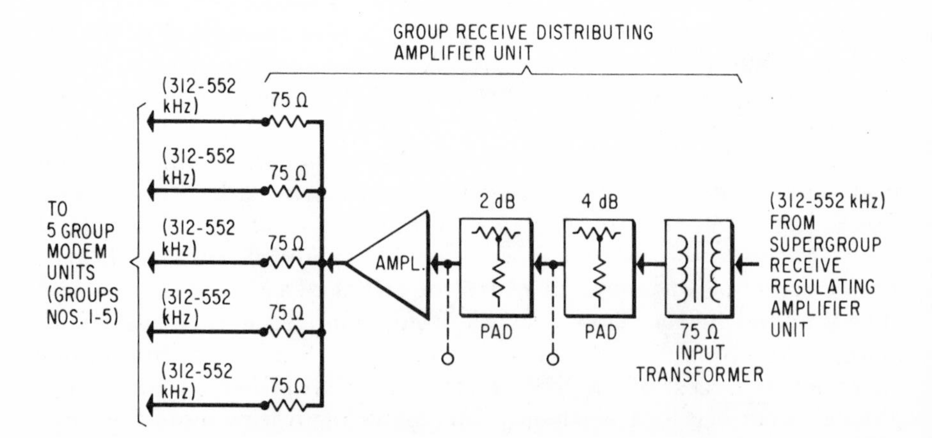

Group Receive Distributing Amplifier Unit (cont.)

drawing. Interactions between the five connected groups are substantially reduced because the distributing amplifier has a zero impedance output. When required, redundancy operation for the group receive distributing amplifier is provided in a manner similar to that of the supergroup receive regulating amplifier unit.

Functions of Receiving Branch of Group Modem Unit

The *receiving branches* of five *group modem units,* as previously explained, are connected in parallel to the output of the group receive distributing amplifier unit. The bandpass filter in each of the group modem receiving branches will select, from the incoming 312-552-kHz band, only the particular 48-kHz frequency segment for which it is designed. The frequency range of the five bandpass filters is listed in the table of the block diagram on the following page. For example, the receive bandpass filter in the modem unit of group No. 1 is 312-360 kHz. For group No. 2, the bandpass filter range is 360-408 kHz, and for group No. 5, it is 504-552 kHz.

The 312-360-kHz signals from the bandpass filter proceed to the group demodulator through a 7-dB ± 3-dB *receive level pad,* as indicated in the drawing. This pad serves to isolate the receive bandpass filter from the group demodulator. It also provides for the final adjustment of the fixed-gain signal level at the output of the receiving branch of the group modem unit.

The group carrier signal applied to the demodulator in the modem unit of group No. 1 is 420 kHz. This is the same frequency that was sent to the modulator in the transmitting branch of this group's modem unit. The demodulation process produces upper sideband frequencies of 732-780 kHz and lower sidebands of 60-108 kHz. The 60-108-kHz lower sideband is selected by the low-pass filter that follows the group demodulator. The upper sideband product (732-780 kHz) is attenuated by the high-pass filter which is also connected to the demodulator's output. This filter, likewise, serves to improve the transmission of the selected group carrier signals by providing a proper impedance termination for the circuit.

The 60-108-kHz-output of the low-pass filter goes to the associated *group receive amplifier unit,* as illustrated in the diagram. The amplifier in this unit has a nominal gain of 42 dB. The output circuit of this amplifier includes a tapped transformer and a 3-dB pad. This circuit arrangement provides for either of two impedance and output signal levels. They are 135 ohms balanced and a level of −5 dBm, or 75 ohms unbalanced and a level of − 8 dBm. If redundancy operation is necessary, a similar redundant amplifier circuit is supplied as described for the supergroup receive regulating amplifier unit.

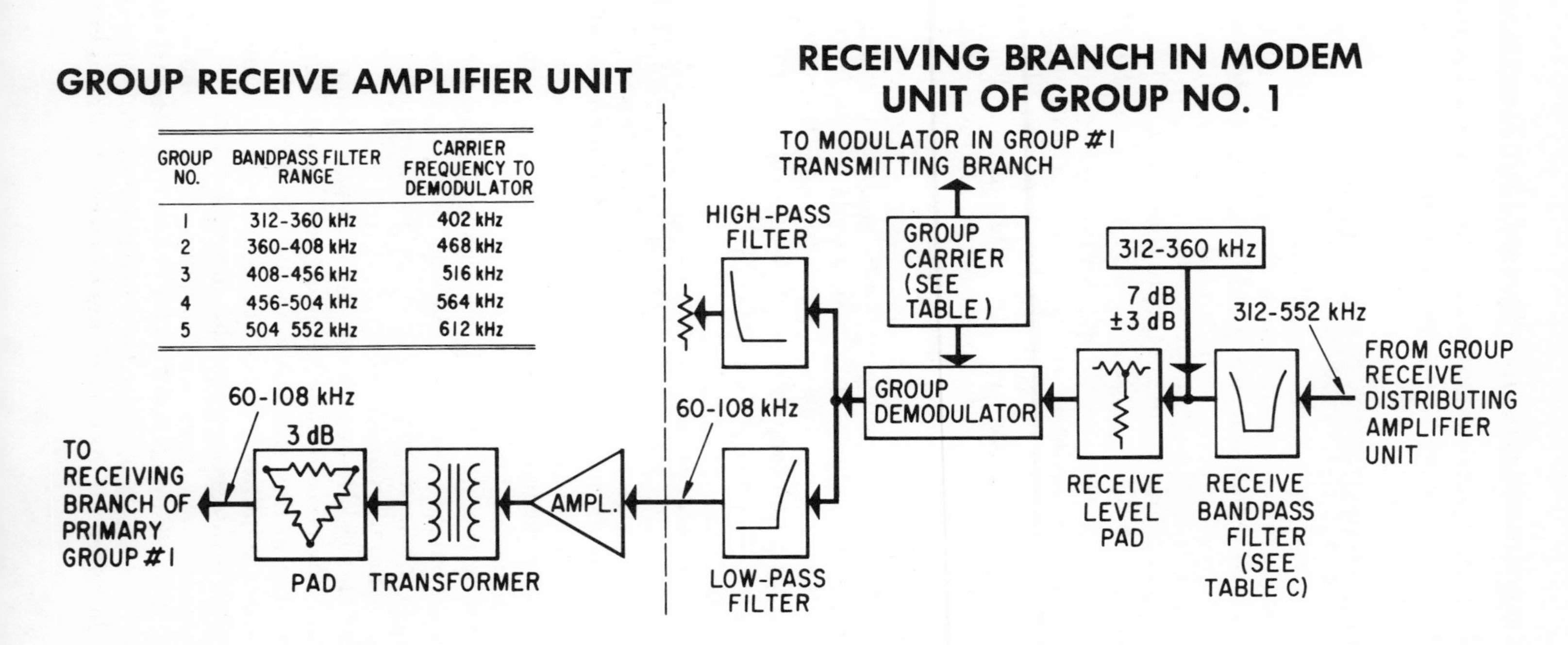

GROUP NO.	BANDPASS FILTER RANGE	CARRIER FREQUENCY TO DEMODULATOR
1	312-360 kHz	402 kHz
2	360-408 kHz	468 kHz
3	408-456 kHz	516 kHz
4	456-504 kHz	564 kHz
5	504 552 kHz	612 kHz

Receiving Branch of Group Modem Unit Functions (cont.)

From the group receive amplifier unit, the 60-108-kHz carrier signals proceed to the *receiving branch of primary group No. 1.*

Receiving Operations in Primary Group

Continuing from the group and supergroup equipment units, the carrier signals proceed to their final destination in the *channel bank* equipment. Note that the 60-108-kHz spectrum, which resulted from the demodulation process in the modem unit of group No. 1, contains 12 VF channels, This frequency band is next directed to the *receiving branch of primary group No. 1,* as shown in the drawing. The 30-dB and 3-dB pads attenuate this incoming carrier signal to a –8-dBm level, for coupling to the demodulator. The 3-dB pad is strapped out if a –5-dBm received level is needed for the demodulation process.

An 8,248-kHz carrier is utilized to demodulate the 60-108-kHz carrier signals. This 8,248-kHz carrier oscillator is also common to the modulator in the transmitting branch of primary group No. 1, as depicted in the block diagram. The products of the demodulation process are 8,140-8,188-kHz lower sideband and 8,308-8,356-kHz upper sideband. These frequency bands, after amplification by the amplifier following the demodulator, are applied to *channel units* No. 1 to No. 12. These twelve channel units, which comprise primary group No. 1, are connected in parallel to the output of the aforementioned amplifier in the receiving branch of primary group No. 1.

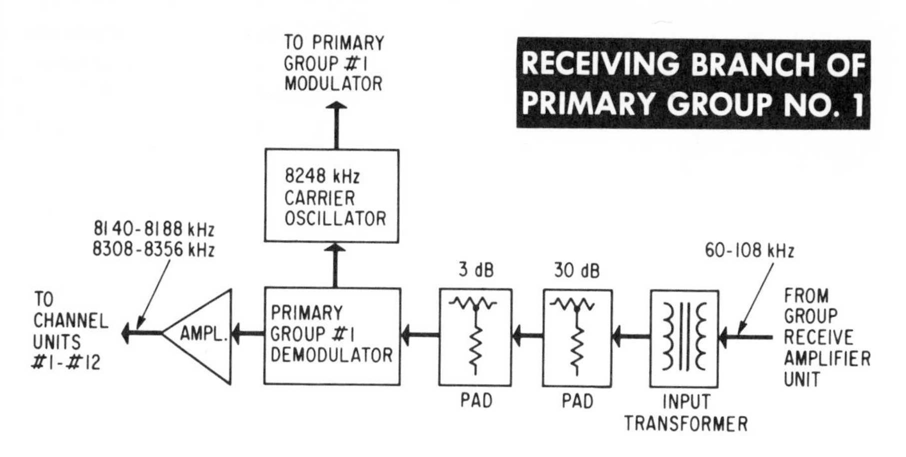

Receiving Operations of Channel Unit

The *receiving branch* of the *channel unit* is the final destination of the received carrier signals. The carrier signals are translated back to voice-frequency signals in the channel unit for transmission to the

Receiving Operations of Channel Unit (cont.)

assigned toll trunk circuit. Recall that we have been following the path taken by carrier signals initially originated in channel No. 4, as explained in the section on Transmitting Operations of Channel Unit. Therefore, we will now refer to the receiving branch circuit of channel unit No. 4, as shown in the block diagram.

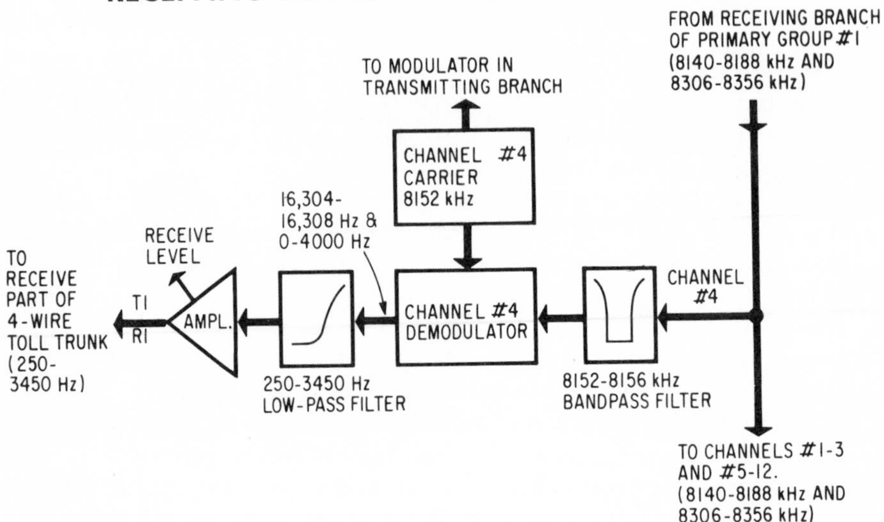

The receiving branches of the twelve channel units comprising primary group No. 1 are connected in parallel, as previously described, to the output of the receiving branch of primary group No. 1. Each of the 12-channel units is equipped with an extremely selective (polylithic crystal type) bandpass filter. Thus, the proper 4-kHz increment of the 8,140-8,188-kHz spectrum is selected for demodulation. Referring to the drawing, channel unit No. 4 is equipped with an 8,152-8,156-kHz bandpass filter. This selected band is demodulated by the 8,152-kHz channel carrier supplied to the demodulator. Channel carrier frequencies for the 12-channel units are in 4-kHz increments in the band from 8,140 kHz to 8,184 kHz. For example, the channel carrier for channel unit No. 1 is 8,140 kHz, for channel unit No. 2 the frequency is 8,144 kHz, and so on to 8,184 kHz for channel unit No. 12.

The products of demodulation in channel unit No. 4 are 0-4,000 Hz lower sideband, and 16,304-16,308 Hz upper sideband. The low-pass filter permits the passage of only the desired VF signals (250-3,450 Hz) which are included in the lower sideband. Other frequencies including the upper sideband are attenuated. The amplifier, whose gain is adjustable, amplifies the VF signals and also provides a 600-ohm balanced 2-wire output termination. This 2-wire *receive output*

Receiving Operations of Channel Unit (cont.)

(T1 and R1 leads) of channel unit No. 4 connects to the associated 4-wire toll trunk circuit. The *receive level* control of the amplifier is used to adjust the VF signal level to the toll trunk circuit in the range from 0 dBm to a maximum of +10 dBm, as may be required.

Questions and Problems

1. Describe the construction of the copper tubes in a coaxial cable. How many copper tubes comprise the coaxial cable placed in a 4-inch (10.2 cm) cable duct? How many carrier systems are handled by this coaxial cable?
2. What was the designation and voice channel capacity of the initial coaxial cable carrier system installed by the Bell System? Why do coaxial cable systems utilize carrier frequencies above 300 kHz?
3. What are the VF channel capacities of the L3, L4, and L5 type carrier systems? Which ones are transistorized? How many voice channels comprise one Mastergroup? One Jumbogroup?
4. Why are carrier alarm and protection circuits so necessary for broadband carrier systems? For microwave radio systems?
5. What factors determine losses in the microwave radio path? What means are generally employed to minimize these losses? What is meant by *free space*?
6. What three microwave bands are usually assigned by the FCC to communication common carriers for high-capacity radio channels? What is the authorized bandwidth for a radio channel in these bands? What type of modulation is employed in microwave radio equipment?
7. How many channel banks comprise a primary group unit in the GTE Lenkurt 46A and Western Electric U600 type multiplex equipment? What is the frequency range of the primary group?
8. What spectrum is occupied by the basic supergroup? How many VF channels are contained in a supergroup? How many supergroups are combined to form a Mastergroup? What frequency band is occupied by the Mastergroup?
9. How many modulation stages are employed in the GTE Lenkurt 46A carrier system to translate the voice signals up to the carrier band that connects to the microwave radio terminal? What is the designation of each modulation circuit involved in the process?
10. What is the function of the Receive Bandpass Filter in the Receiving Branch of the Group Modem Unit? What is the bandpass filter range and the demodulator's carrier frequency of Group No. 5?

Index